NIGHTWATCH

AN EQUINOX GUIDE TO VIEWING THE UNIVERSE

REVISED AND UPDATED

TERENCE DICKINSON

ILLUSTRATIONS BY VICTOR COSTANZO AND ADOLF SCHALLER

FEATURING MORE THAN 70 PHOTOGRAPHS BY AMATEUR ASTRONOMERS

FIREFLY BOOKS

ACKNOWLEDGMENTS
Thanks to Roy Bishop, Glenn Chaple, Alan Dyer, Leo Enright and David Rodger, who carefully read the manuscript for the first edition and offered useful comments. Special thanks to James Lawrence, who wanted it done the right way from the start, to Pamela Wimbush, graphic designer of the first edition, and to Roberta Voteary, designer of this new edition. Ray Villard of the Space Telescope Science Institute, Baltimore, Maryland, provided custom photographs of planetarium sky projections for the 20 detailed charts. The illustrations and charts by Victor Costanzo and Adolf Schaller have stood the test of time and remain what I regard as the best of their type ever published in an introductory astronomy book.

Second printing 1984
Third printing 1985 (revised)
Fourth printing 1986
Fifth printing 1987
Sixth printing 1988
Seventh printing 1989 (major revision)
Eighth printing 1992 (updated)
Ninth printing 1993 (updated)
Tenth printing 1994
Eleventh printing 1995
Twelfth printing 1996
Thirteenth printing 1997

Canadian Cataloguing in Publication Data

Dickinson, Terence
NightWatch

Includes bibliographical references and index.
ISBN 0-920656-91-9 (bound)   ISBN 0-920656-89-7 (pbk.)

1. Astronomy—Popular works.   I. Title.

QB64.D53 1992   520   C92-093750-0

Published by: Firefly Books Ltd., 3680 Victoria Park Avenue, Willowdale, Ontario, Canada M2H 3K1

Published in the U.S. by: Firefly Books (U.S.) Inc., P.O. Box 1338, Ellicott Station, Buffalo, New York 14205

Printed and bound in Canada by: Friesens, Altona, Manitoba

Front Cover: Two-photograph composite: observers using a 14-inch telescope, by Alan Dyer; and time-exposure of star field in Cygnus, by Michael Watson.

Inside Front Cover: Open cluster M35 (upper left) and open cluster NGC 2158 (lower right), photographed by Evered Kreimer, 12-inch Newtonian telescope.

Preface Illustration: Orion Nebula, photographed by Jim Riffle with a 24-inch Astro Works Schmidt-Cassegrain telescope.

Back Cover: Aurora of March 12-13, 1989, photographed from eastern Ontario by Terence Dickinson. Overexposed moon is at lower left.

To Susan, who shares the celestial voyage

ALSO BY TERENCE DICKINSON:

The Universe and Beyond
Exploring the Night Sky
Exploring the Sky by Day
The Backyard Astronomer's Guide (with Alan Dyer)
From the Big Bang to Planet X
Extraterrestrials
Other Worlds
Summer Stargazing

Contents

Preface

In the years since the first edition of *NightWatch* appeared in November 1983, more than 100,000 copies have found their way into the hands of astronomy enthusiasts in Canada and the United States. For me, the most gratifying part of this successful publishing story has been the letters and comments I have received from hundreds of backyard astronomers who have told me how the book helped them through the crucial initial stages of celestial exploration. The overriding goal in earlier editions of *NightWatch* and in this extensively revised and slightly expanded version has been to provide a *complete* first book on amateur astronomy.

In this major revision, I have not tampered with the basic structure and presentation, but extensive fine tuning and updating have touched almost every page. The most visible of the changes is a replacement of all professional observatory photographs with amateur photographs. Amateur colour astrophotography has now reached a technical plateau where it is surpassing all but the very finest professional efforts. The work of Tony Hallas, Michael Watson, Jim Riffle, Jack Newton, Alan Dyer and many others listed on page 160 is a pleasure to behold, and I am delighted that *NightWatch* can be a gallery for such splendid pictures. (Five spacecraft images in Chapters 1 and 7 are the only photographs in the book that are not the result of amateur work.)

Other changes include almost complete rewrites of Chapters 5 and 11 due to evolution in amateur telescope equipment and advances in films suitable for astrophotography. Tables have been updated throughout. Many other improvements are the direct result of suggestions from readers. Further comments are always welcome.

In response to readers' comments and requests, a companion volume to *NightWatch* is now available. The new book, *The Backyard Astronomer's Guide*, published by Camden House, provides more detailed practical information on techniques of celestial observation and in-depth material on the selection and use of astronomical equipment. The same generous use of illustrations and the step-by-step nontechnical approach used in *NightWatch* are retained in the larger companion volume.

Although more people than ever before are dabbling in amateur astronomy and the range and quality of equipment to pursue the hobby have never been better, there is a dark cloud on the stargazing horizon. An ever-growing menace is threatening the night we cherish. The foe is light pollution — streetlamps, outdoor-sign illumination, parking-lot lights, building lights and outdoor fixtures around private homes. Combined, these sources have created great domes of light over our cities that have beaten back the stars. The glow is visibly growing every year. Anyone who cares to see the natural beauty of a dark night sky has to flee ever farther into the country. An evening of stargazing has become an expedition for many aficionados.

Nighttime lighting has expanded much faster than has the population. By one estimate, it has quadrupled in overall intensity every decade since 1960. I am not suggesting that people cannot have lighting at night if they want it, but with present lamp designs, about one-quarter of all outdoor-light output floods uselessly sideways and upward, doing nothing but destroying our heritage of a naturally star-filled night. Awareness of this waste, which costs roughly a billion dollars annually in Canada and the United States, is the only way to begin to bring it under control.

Stargazers, amateur astronomers, astronomy enthusiasts — whatever we call ourselves — are naturalists of the night. We appreciate what so many ignore. We look to the night sky as a wonderland for exploration, and we notice it being tarnished. But awareness is growing. In the 21st century, dark-sky preservation will be a necessary and accepted form of environmental protection.

— Terence Dickinson
Yarker, Ontario

1

Discovering the Cosmos

We are voyagers on the Earth through space, as passengers on a ship, and many of us have never thought of any part of the vessel but the cabin where we are quartered.

– S.P. Langley

Imagine a world where a thimbleful of matter weighs as much as Mount Everest. The gravitational pull at the surface of this celestial body is so enormous that a human visitor would be crushed instantly by his or her own weight into a puddle no thicker than an atomic nucleus.

Now consider a planet whose sky is ablaze with the fireworks of stars being ripped into gaseous tendrils by a bizarre gravity funnel. There can be no living creatures in this corner of the universe, for the region is perpetually bathed in ultralethal doses of x-ray radiation.

Envision a planet cloaked in a choking blanket of carbon dioxide laced with sulphuric-acid rain. Here, it is so hot that lead would be a convenient liquid to use in a thermometer. A human explorer would be simultaneously incinerated and asphyxiated in the hellish environment.

Then there is a realm where two suns illuminate the cosmic landscapes. The solar twins waltz in a deadly gravitational embrace, with millions of tons of gaseous star-stuff flowing between them every second. The tug-of-war will end in the explosive destruction of one of the stars, reducing any nearby worlds to cinders.

All these alien vistas exist. The dense supergravity object is the collapsed core of an exploded star, known as a neutron star. The gravity funnel is a massive black hole that is believed to be lurking at the centre of the Milky Way Galaxy. The carbon-dioxide hothouse is Venus, nearest planet to Earth. The twin-sun system is known as Beta in the constellation Lyra, seen overhead on midsummer evenings. In a universe of a billion trillion stars and an unknown number of planets, the imagination withers in any attempt to grasp the true diversity of the cosmos.

Yet for me, as I stand under the nighttime canopy of stars, that has always been the lure of the backyard exploration of the universe: the chilling realization that Earth is but a mote of dust adrift in the ocean of space. The fact that Earth harbours creatures who are able to contemplate their place in the cosmic scheme must make our dust speck at least a little special. But wondering who else is out there only deepens the almost mystical enchantment of those remote celestial orbs.

Since I first became fascinated with the cosmos over 30 years ago, humankind's knowledge of the universe has expanded enormously. It has been almost a continuous intoxicant for me as the discoveries have come thick and fast: quasars, pulsars, black holes, volcanic moons, bodies made entirely of ice orbiting larger worlds of liquid hydrogen. Hardly a month goes by without something new to ponder.

The mental exercise of grappling with the vast distances and sizes of celestial objects is captivating in itself. But to be able to stand under those stars and planets on a dark night is what makes backyard astronomy an addictive pastime. For me, it is communing with nature on a grand scale. I have come to know those remote stars and galaxies. The stellar panorama comes alive when one can recall: "There's

the star that is 250 times bigger than our sun . . . and over there, in a spot I can cover with my fingernail, is a cluster of 500 galaxies, each like the Milky Way . . . and over here is the nucleus of our galaxy, just behind that rift in the Milky Way." All this can be seen and appreciated with the unaided eye. The experience is both humbling and exhilarating. From our tiny island planet, a huge sector of the universe is visible.

For thousands of astronomy enthusiasts, the ultimate in self-discovery is the exploration of some of these celestial wonders with a backyard telescope. My telescope has shown me the galactic nurseries where stars are born and the gaseous tombstones they leave behind when they die. One night not long ago, my telescope, which is typical of those used by amateur astronomers, revealed the delicate spindle-shaped image of a galaxy 70 million light-years away. The light that I saw from that remote continent of stars had been hurtling through space at 670 million miles per hour since the time dinosaurs ruled Earth. Backyard astronomy is just as much a cerebral as a visual adventure.

Little more than a generation ago, an enthusiast had to be mechanically skilled to own a telescope. Few astronomy buffs could afford commercially made telescopes, which were generally handcrafted and very expensive, so most made their own instruments. Today, mass-produced telescopes of good quality are available for less than a thousand dollars. Five thousand will buy a telescope that can outperform virtually anything used by astronomy hobbyists a generation ago.

When viewing the moon through a top-quality 6- or 8-inch telescope (the sizes often used by amateur astronomers today), one can detect craters the width of a large football stadium and subtle ripples only tens of feet in height on the floors of the lunar plains — views corresponding to those seen out the window of a spacecraft only a few hundred miles above the moon's surface.

When you turn a typical amateur telescope to Jupiter, the giant planet's ocean of clouds exhibits belts and zones in yellow, salmon, grey, white and brown. Jupiter's four large moons circle the giant like obedient servants dithering to please the great master. Then there is Saturn's exquisite system of rings casting plainly evident shadows on the great planet.

Beyond the solar system, a limitless hunting ground awaits. A telescope, or even a pair of binoculars, transforms the gauzy Milky Way into a glittering array of individual stars. Elsewhere, the differing colours of the stars are on display, from pulsating variables that occasionally shine with a blood-red hue to double-star systems, where one sun may be orange and the other blue.

Star clusters ranging from loose collections of 10 or 20 suns to colossal swarms of hundreds of thousands of stars are out there if you know where to look. And if one probes even deeper into space, thousands of galaxies — star cities beyond our own Milky Way system — dot the void like puffs of frosty breath frozen in time.

Although this celestial showcase has always been there for anyone who has access to a good telescope, only recently have large numbers of such telescopes been in the hands of amateur astronomers. This book is designed primarily for those with a budding interest in the cosmos who have yet to purchase a telescope. By untangling the jargon and avoiding what I have found is unnecessary technical baggage, this book will enable the uninitiated to make intelligent decisions about pursuing backyard astronomy and purchasing a telescope that will provide a lifetime of celestial excursions.

But the book does not overlook the surprising variety of celestial phenomena revealed to the unaided eye: the diaphanous auroral curtains that sometimes drape the northern sky, the dance of two planets as they slip past one another in their paths around the sun, another galaxy of stars like the Milky Way two million light-years away, and more. Exploring the night sky is, in many ways, like a sightseeing tour of exotic foreign lands. But, as with any journey, the appreciation of that tour is greatly enhanced when the tourist has prepared for the venture. Once experienced under the right conditions, the cosmic panorama will tempt again and again.

Above left, *the moon and Venus, a striking duo at dawn.* **Above right**, *the Earth's nearest celestial neighbour poses over a terrestrial outcrop; the reverse view,* **lower right**, *captured by the Apollo 17 astronauts, reveals the stark contrast between our planet and the lifeless lunar surface.* **Lower left**, *scanning the skies.*

2

The Universe in Eleven Steps

The universe begins to look more like a great thought than a great machine.

— Sir James Jeans

Astronomy stretches the mind like nothing else in human experience. Enormous distances, sizes and time spans challenge the imagination; alien concepts like black holes and galactic cannibalism strain comprehension. Yet it is possible to bring the total picture — the structure and extent of the entire universe — into focus.

When our grandparents were children, no one knew what, if anything, existed beyond the stars. The discovery of the essential components of the universe has been largely a 20th-century enterprise. The accelerated pace of discovery in recent years has given astronomers reason to believe that there *is* an end to space and time — the universe is finite. It would be ideal to be able to show the various components of the entire universe in a single illustration, but that would require a whole wall. The only way to achieve a realistic perspective in book form is through a series of ever-widening views.

The illustrations in this chapter, rendered by astronomy artist Adolf Schaller, are somewhat more than an "artist's conception." They are scientifically rig-

STEP 1: *The first of the eleven steps places Earth in an imaginary cube only slightly larger than the planet itself. Earth is 7,915 miles (12,742 km) in diameter; its mass equals 6 billion trillion tons. Yet it is but a mote of dust in the universe, as the widening perspectives on the following pages reveal.*

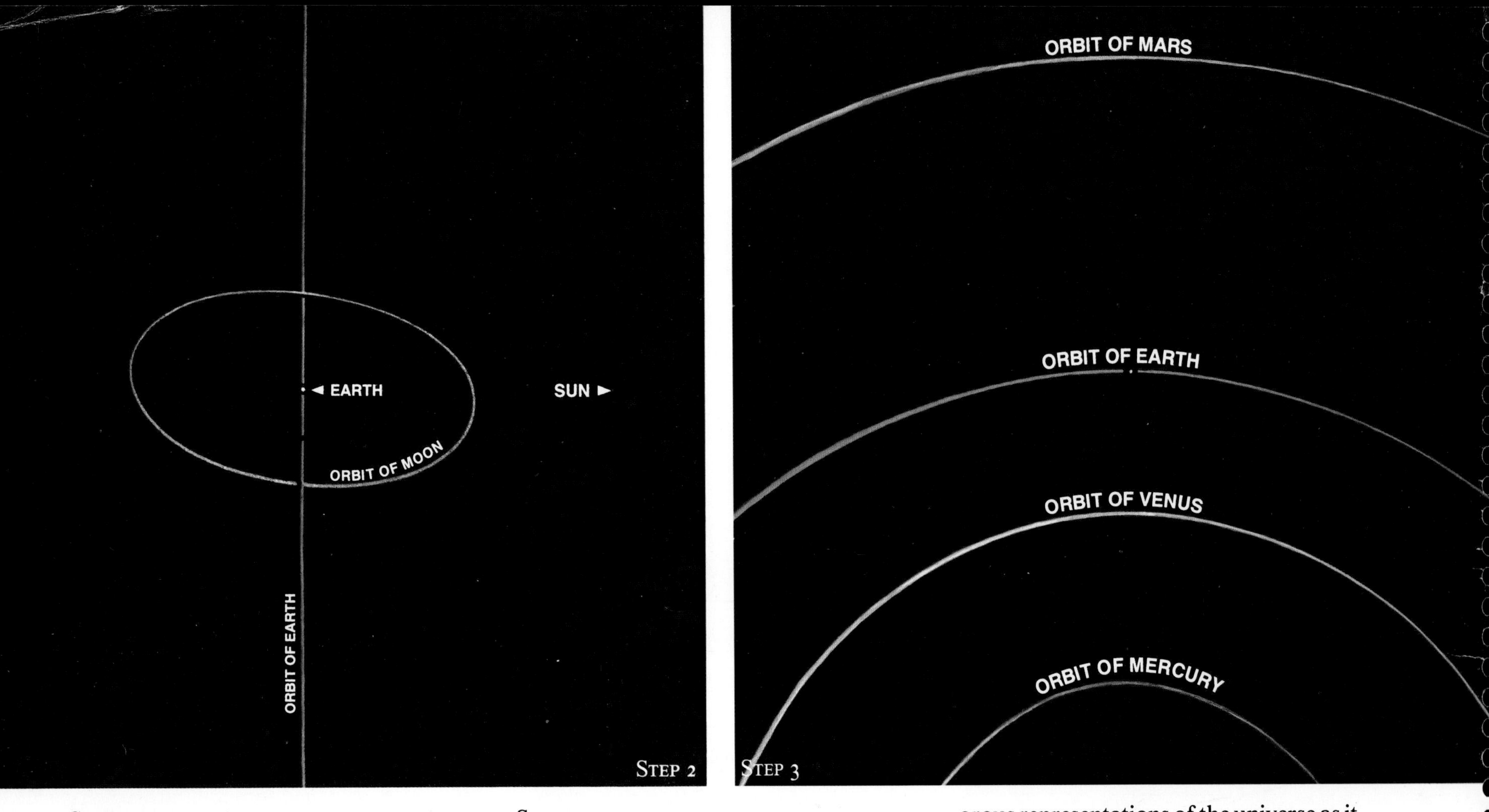

STEP 2

Width of cube = 1.5 million km, or 932,000 miles
Light time across cube = 5 seconds
Volume of cube = 3.38 million trillion cubic km, or 810 million billion cubic miles
Distance from Earth to moon = 384,500 km, or 239,000 miles
Orbital period of moon = 27.32 days
Diameter of moon = 3,476 km, or 2,160 miles
Mass of moon = 1.2% of Earth
Volume of moon = 1.6% of Earth

STEP 3

Width of cube = 1 astronomical unit, or 150 million km, or 93 million miles
Light time across cube = 8.3 minutes
Volume of cube = 1 cubic astronomical unit
Minimum distance from Earth to Venus's orbit = 0.27 astronomical unit
Minimum distance from Earth to Mars' orbit = 0.38 astronomical unit
Minimum distance from Earth to Mercury's orbit = 0.53 astronomical unit
Orbital period of Mercury = 88 days
Orbital period of Venus = 224¾ days
Orbital period of Earth = 365¼ days
Orbital period of Mars = 687 days

STEP 4

Width of cube = 100 astronomical units
Light time across cube = 14 hours
Volume of cube = 1 million cubic astronomical units
Orbital period of Jupiter = 11.86 years
Orbital period of Saturn = 29.46 years
Orbital period of Uranus = 84.0 years
Orbital period of Neptune = 164.8 years
Orbital period of Pluto = 248 years
Distance from Sun to: Mercury 0.39 astronomical units; Venus 0.72; Earth 1.00; Mars 1.52; Jupiter 5.20; Saturn 9.54; Uranus 19.2; Neptune 30.1; Pluto 24.6 to 52.6*

*Pluto's elliptical orbit has placed the planet closer to the sun than Neptune from 1979 to 1999.

STEP 5

Width of cube = 10,000 astronomical units, or 0.16 light-year
Light time across cube = 58 days
Volume of cube = 1 trillion cubic astronomical units, or 0.004 cubic light-year
Estimated number of comets inside (or just beyond) cube = 100 billion
Estimated diameter of average comet = 2 km, or about 1 mile
Estimated total mass of all known comets = 100 times mass of Earth

orous representations of the universe as it is presently known. Each drawing represents a cubic volume of space a million times larger than the one before. That is, in each increment, the sides are extended 100 times, although the cube remains centred on Earth.

We begin by enclosing Earth in an imaginary box whose sides are slightly larger than the diameter of the planet. This is a manageable starting point. Earth has dimensions that can be easily grasped. Many people routinely fly a quarter of the way around the globe in a few hours. In a high-flying jet liner, it is possible to get a hint of the Earth's curvature at the horizon. Above, the deep blue signals the thinning atmosphere at the edge of space.

The next step outward reduces Earth to a dot, since each side of the cube has expanded 100 times to just less than a million miles. The cube now comfortably contains the moon's orbit, a half-million-mile-diameter circuit around Earth. The distance from Earth to moon is almost exactly 30 Earth diameters, a modest leap

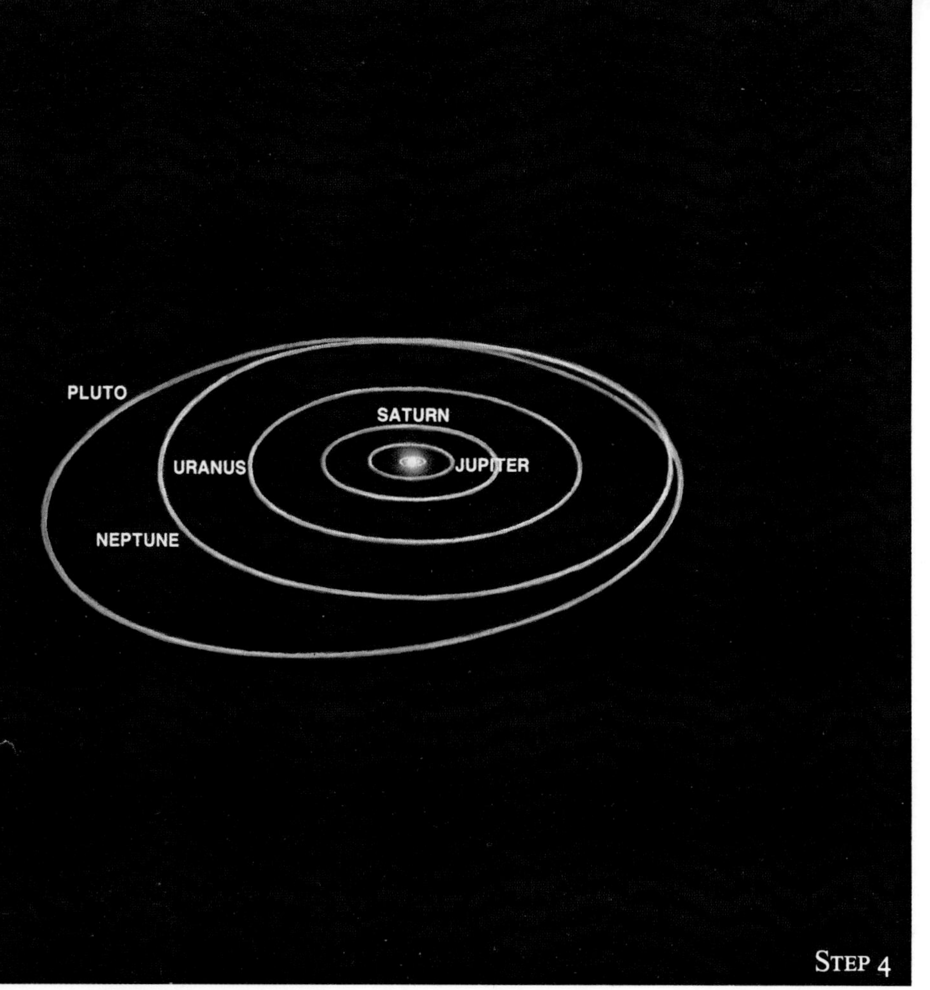

STEP 4

SOLAR SYSTEM

STEP 5

that required two days for the Apollo astronauts to span. The fastest interplanetary spacecraft could traverse it in about six hours. No other space objects, except artificial satellites, are within the bounds of this second cube.

The third cube, slightly less than 100 million miles across, happens to coincide with the dimension of a standard cosmic measuring unit known as the astronomical unit, or AU. One AU is the Earth-sun distance, 93 million miles, or 150 million kilometres. Centred on Earth, this cube includes portions of the orbits of Mercury, Venus and Mars, our nearest planetary neighbours. A score of Soviet and American robot spacecraft have navigated the gulf between these worlds. Travel times are a minimum of four months.

The region between planets is essentially empty space, except for a few renegade asteroids — chunks of rocky material left over from the formation of the planets — that have drifted in from the asteroid belt beyond Mars. The largest of these are flying mountains that pose the threat of mass destruction if one hits Earth. Fortunately, such collisions are exceedingly rare. The last big impact is thought to have occurred about 65 million years ago, around the time of the extinction of the dinosaurs.

Other intruders in our basically quiet neighbourhood are comets, also mountain-sized but largely composed of ice. When melted by solar radiation, comets sprout the familiar filmy tails seen in photographs. Apart from these visitors from other parts of the solar system, our sector of space is tranquil and uneventful. It must have been this way for some time: Earth has orbited the sun nearly five billion times since it formed and is still here.

The Earth's companion planets, endlessly swinging their gravitationally prescribed paths about the sun, are a varied lot ranging from crater-pocked moonlike Mercury to colossal Jupiter, a globe of liquid hydrogen whose bulk equals 1,000 Earths. If there is one overriding lesson that has been learned from the interplanetary spacecraft probes, it is that those other worlds are more alien than the most vividly creative science fiction writer ever imagined.

To include all nine planets in the solar system out to Pluto, we take the fourth increment in our progression — a cube 100 AU wide. The only solar system members left outside this cube are most of the comets, whose elongated orbits usually keep them up to a trillion or more miles from the sun. If viewed from this far out in space, all the planets would be tiny specks, invisible to the unaided eye. (We have exaggerated a little by showing them and their orbits.) On this scale, the blazing sun is the only object of significance; the planets, by comparison, are merely debris orbiting around it.

When astronomers look at other stars from Earth, even using the most powerful telescopes, no planets can be seen. Like the sun, the stars completely overpower their planetary systems, masking even worlds as great as Jupiter. However, in 1983, a vast ring of billions of asteroids was detected orbiting the star Vega, add-

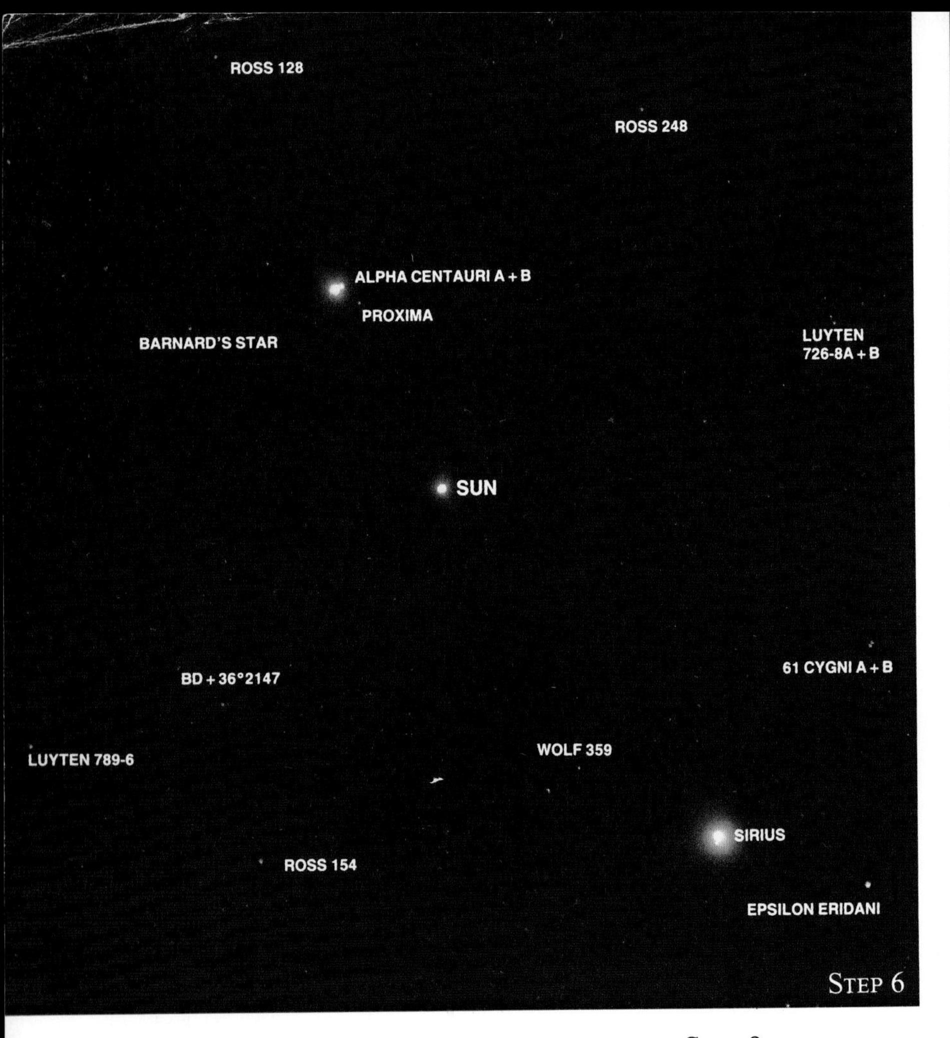

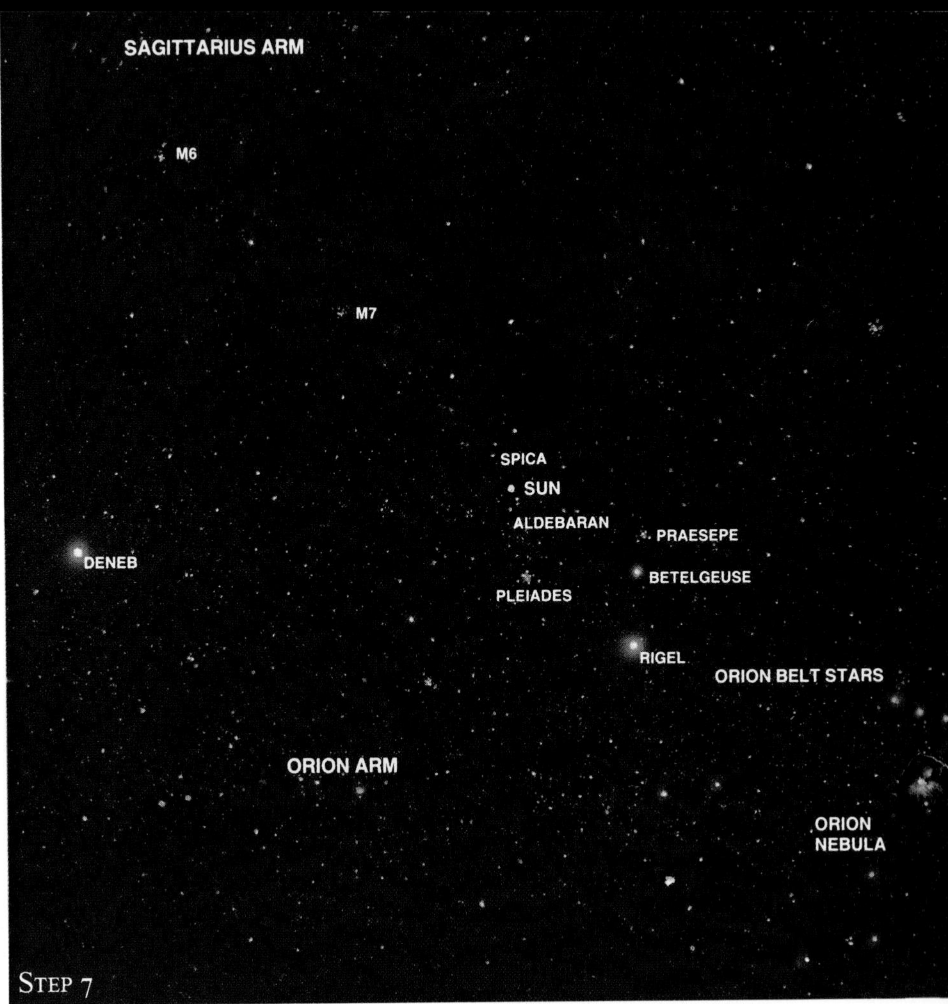

STEP 6

Width of cube = 1 million astronomical units, or 16 light-years
Light time across cube = 16 years
Volume of cube = 4,096 cubic light-years
Number of stars in cube = 17 (8 single stars; 3 double systems; 1 triple)
Actual brightness of some nearby stars compared to sun: Alpha Centauri A 1.2; Alpha Centauri B 0.36; Proxima Centauri 0.00006; Barnard's star 0.00044; Wolf359 0.00002; BD + 36°2147 0.0052; Sirius 23; Epsilon Eridani 0.3; 61 Cygni A 0.08; 61 Cygni B 0.04

STEP 7

Width of cube = 1,600 light-years, or 490 parsecs
Volume of cube = 4 billion cubic light-years
Estimated number of stars in cube = 2 million
Distance from sun to: Aldebaran 68 light-years; Spica 220; Pleiades 410; Betelgeuse 520; M7 cluster 800; Deneb 1,600; Orion Nebula 1,600

STEP 8

Width of cube = 160,000 light-years, or 49 kiloparsecs
Volume of cube = 4,000 trillion cubic light-years
Estimated number of stars in cube = 1 trillion
Diameter of Milky Way Galaxy = 100,000 light-years
Orbital period of sun around galaxy = 220,000 years
Mass of LMC = 10 billion solar masses
Mass of SMC = 2 billion solar masses

STEP 9

Width of cube = 16 million light-years, or 4.9 megaparsecs
Number of major spiral galaxies in cube = 6
Estimated number of minor spirals and dwarf galaxies = 100
Estimated number of stars in cube = 10 trillion

ing to numerous pieces of indirect evidence that other planets exist. Their reality will likely be confirmed before the end of the century.

The fifth step outward produces a 10,000-AU cube, enclosing almost exclusively empty space. Pluto's orbit has shrunk to a tiny oval near the central sun. From the edge of the cube, the sun would appear as simply a very bright star. A haze of comets known as the Oort cloud — named after the Dutch astronomer who first suggested that billions of comets roam the fringes of the solar system — is emphasized for clarity in the illustration. The nearest star is 50 times farther away from the sun than the edge of this cube.

From now on, miles or kilometres become useless for scaling distances. Even astronomical units will soon be cumbersome. For tabulating interstellar distances, astronomers use light-years — the distance that light travels in a year at its constant velocity of 7.2 AU per hour. One light-year is 63,240 AU, making the one-million-AU cube, our sixth increase, about

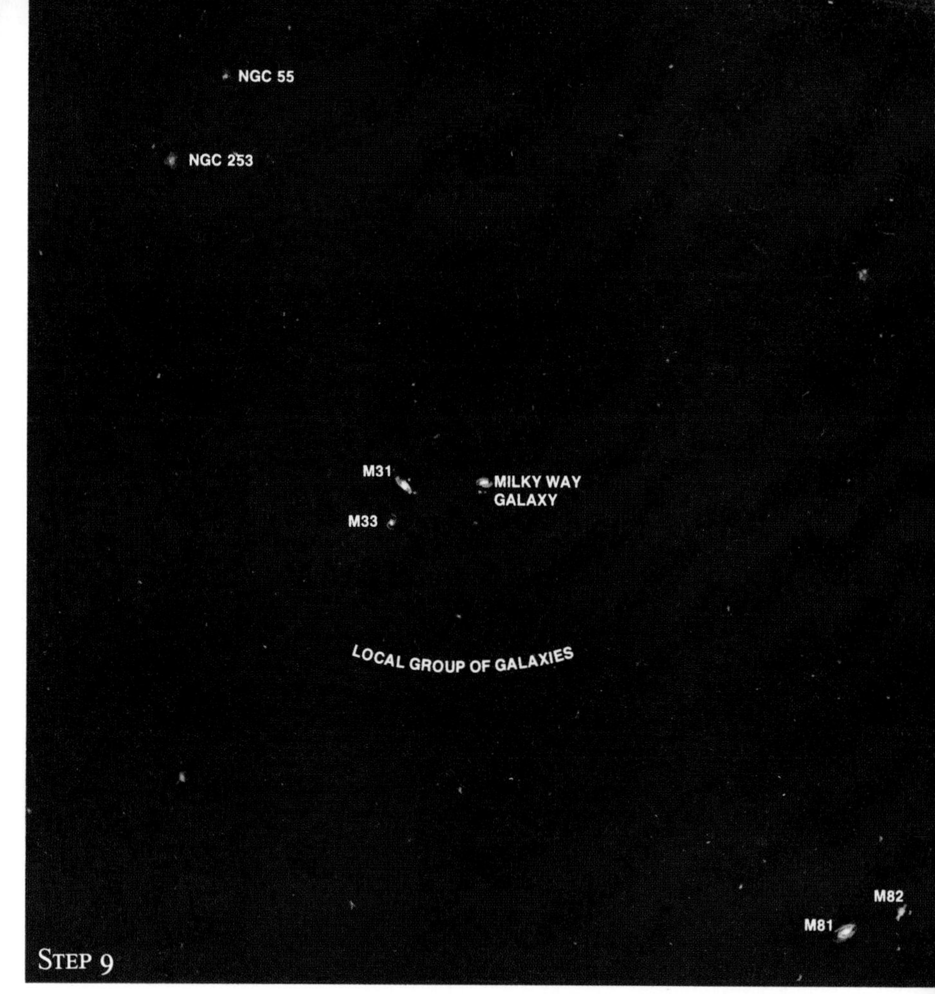

16 light-years wide, or 4.9 parsecs, using another astronomy yardstick.

The sun now takes its place as one star among many. Our nearest neighbour, Alpha Centauri, is a triple-star system 8,000 times farther from Earth than frigid Pluto. Distances between star systems are awesome. If the sun were the size of a grapefruit, this cube would have to be larger than Earth to provide the correct scale.

Comparison of our neighbour stars reveals that the sun is much brighter than the average star. The majority of stars are dim suns called red dwarfs. Only five percent of all stars are equal to or brighter than our sun. Sirius, the brightest star in the night sky, is also the nearest naked-eye star visible from midnorthern latitudes. Because of its celestial location not far from the south-sky pole, Alpha Centauri is seen only from Miami or farther south.

Each star in the sun's vicinity has about 400 cubic light-years of empty space around it, plenty of room for Captain Kirk and the crew of the starship *Enterprise* to scoot around without finding anything. Or would they? Apart from the occasional vagabond comet at the outskirts of each star's Oort cloud, is there anything there? The space between stars is not a total vacuum, but it is close. There is only one atom for every cubic inch, compared with 100 million trillion atoms in a cubic inch of air at sea level on Earth. The only substantial objects that might exist undetected between the stars are black holes, gravity whirlpools believed to be created when certain types of massive stars explode. If black holes are out there, they could not be much closer to us than the stars. Astronomers say that a black hole as massive as the sun, one or two light-years away, could be detected by its gravitational influence, however slight, which would tend to speed up or slow down some of the planets as they ply their orbits about the sun.

The seventh cube in our outward march, 1,600 light-years across, encloses about two million stars, several hundred times more than are visible with the unaided eye on the clearest of nights. The stars seem to crowd on top of one another, but this is only due to the problem of having to show so much in such a small space. Remember that they are all several hundred thousand AU from each other. A few prominent naked-eye objects are identified in this seventh box, including some bright stars and the nebula in the constellation Orion as well as the Pleiades star cluster in Taurus. All are several hundred light-years distant. Although the vast majority of the stars seen on a typical clear night are contained in this celestial cube, it represents only about one-thousandth of one percent of our star city, the Milky Way Galaxy.

The eighth cube, 160,000 light-years wide, spans the Milky Way Galaxy, which is believed to be slightly more than 100,000 light-years in diameter. The Milky Way may contain as many as one trillion stars. There are at least 50 times as many stars in our galaxy as there are humans on Earth.

In the sun's vicinity, roughly two-thirds

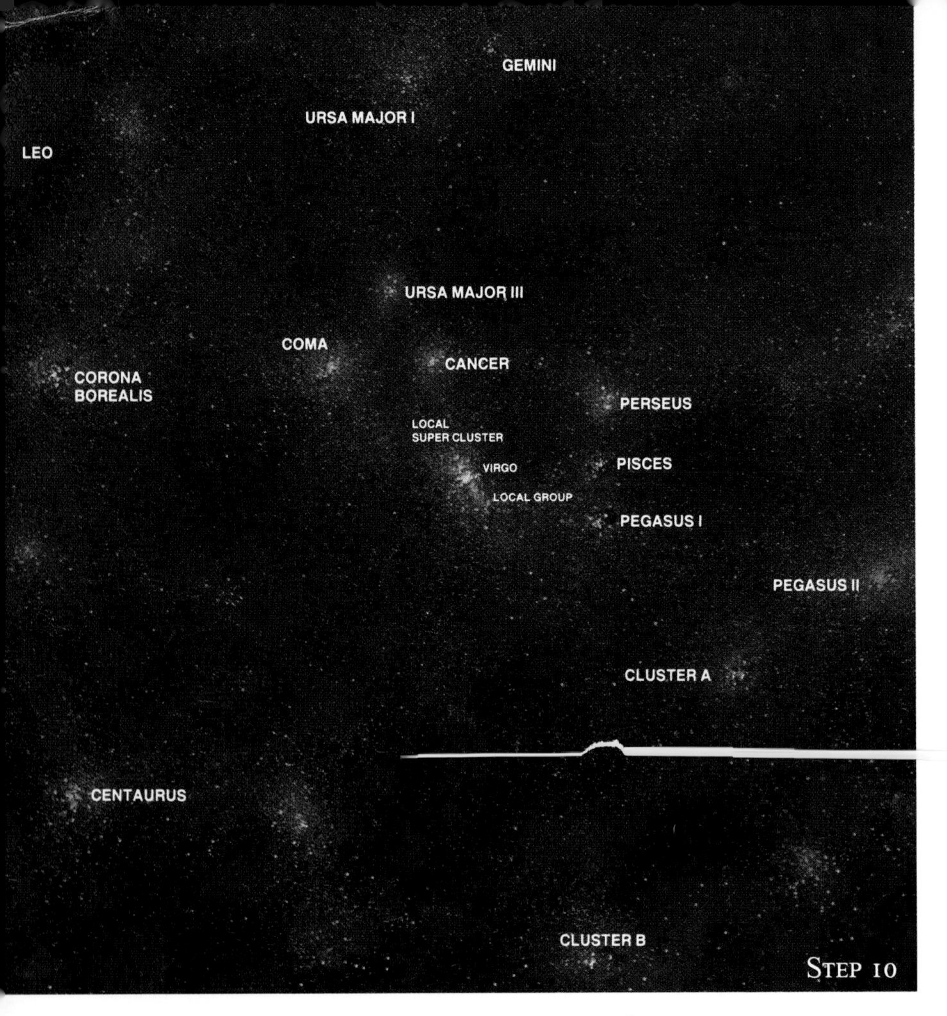

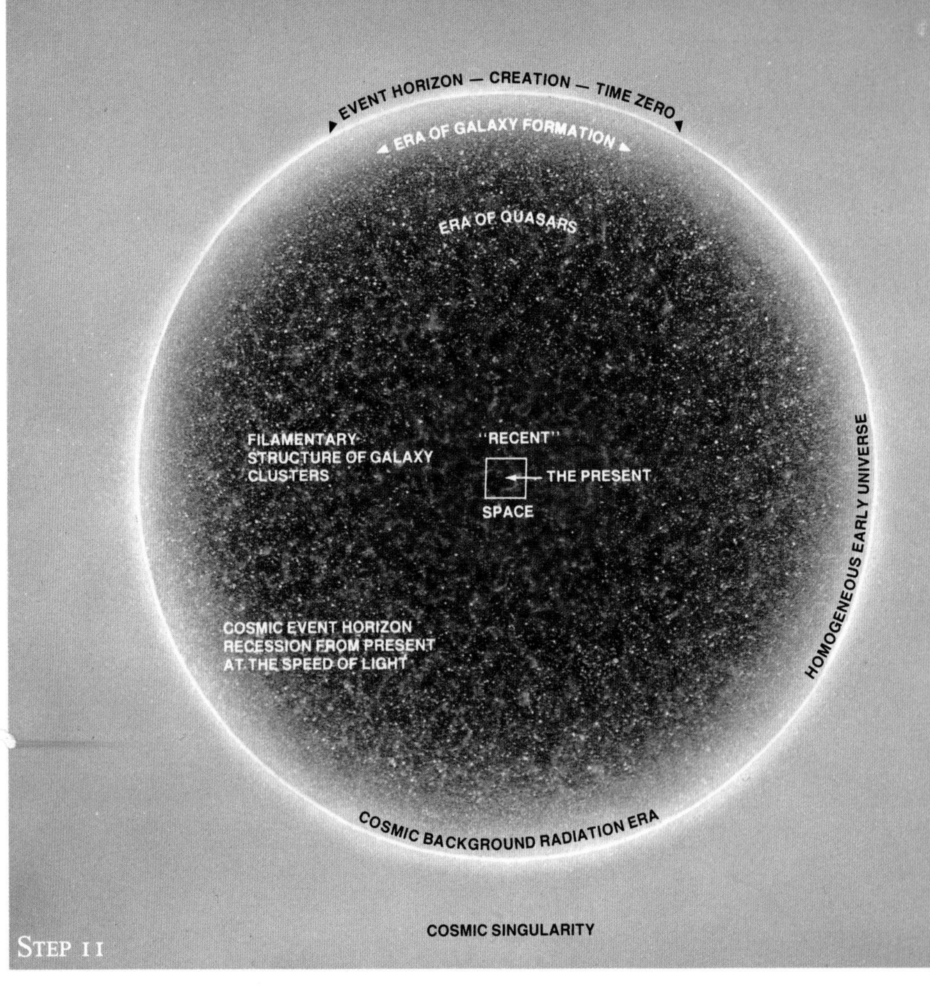

STEP 10

Width of cube = 1.6 billion light-years, or 490 megaparsecs
Estimated number of galaxies in cube = several million
Estimated number of stars in cube = several thousand trillion
Average distance between galaxy superclusters = 300 million light-years

STEP 11

Width of cube = greater than the known universe
Estimated number of galaxies in universe = 100 billion
Estimated number of stars in universe = at least 1 billion trillion
Estimated number of planets in universe = very uncertain, perhaps trillions
Estimated age of universe = 13 to 20 billion years
Evolutionary destiny of universe = unknown

the distance from the hub to the edge, the galaxy is about 3,000 light-years thick. For the proper thickness-to-width ratio, imagine two long-playing records held together. Our sun is like a speck of dust between the records. Most of the stars near the sun move in basically circular paths around the hub — along the record grooves. Moving at a velocity of one AU per week, it takes the sun about 220 million years to complete one revolution around the galactic nucleus. Since the sun and Earth were born, they have made fewer than 25 such trips.

The ninth increment in our push to the edge of the universe reveals a cube 16 million light-years on a side. Now, our galaxy is simply one of several dozen speckling an enormous volume of space. The galaxies within three million light-years of the Milky Way are gravitationally bound into a permanent family. Astronomers call this collection of galaxies the Local Group. Only one of these, the Andromeda Galaxy, is comparable in size to our own. The rest are less than one-twentieth as massive.

Beyond the Local Group, near the limits of the cube, other galaxies similar to the Milky Way hint at what lies in the depths of space.

The space between galaxies is as close to a complete vacuum as can be imagined. Only one atom per cubic metre rides the intergalactic void. Long before Earth existed, matter in the universe clumped into colossal islands that eventually became galaxies. They range from monsters 100 times more massive and several times larger than the Milky Way Galaxy to midgets containing only a few thousand stars. The galaxies are nature's greatest building blocks. So far as astronomers can surmise, a few billion of them constitute the known universe.

To see how galaxies form the fabric of the universe, we move on to the tenth step, a cube of space nearly two billion light-years across. Within this volume, millions of galaxies swim in a seemingly endless abyss. The Milky Way is lost among the galactic throng. But there is structure. Visible at last is the grand architecture of the cosmos. The galaxies are arranged not at random but in knots and diffuse clumps. These superclusters, as they are called, are collections of smaller galaxy clusters like the Local Group. Some superclusters contain tens of thousands of galaxies.

The supercluster to which we belong, the Virgo galaxy supercluster, has at least 3,000 galaxy members and is about 100 million light-years across. The Local Group is near the outer edge. The galaxies M81 and M82, on the right side of the previous cube, are also members, somewhat nearer the supercluster's central zone.

The reason that galaxies are arrayed in tattered but discrete clusters, rather than being haphazardly dispersed, is possibly due to a clumping of the stuff from which the galaxies emerged. One theory suggests that soon after the universe formed, its matter swirled into ribbon- and pancake-shaped slabs which later became the birthplaces of the galaxies. Such speculation is on the cutting edge of astronomical research, but its basic premise depends on the now well-accepted thesis that the universe is finite.

There is some fairly persuasive evidence that our universe does not go on forever. Current estimates suggest that the edge is about 15 billion light-years beyond the limits of this supercluster cube. We now take the audacious step of one final extension in our cosmic progression, this time by a linear factor of 20, to enclose the entire universe.

The suggestion that the universe really ends somewhere is based on the idea that the universe has existed for a finite period of time. According to the widely accepted Big Bang theory, the universe was created about 15 billion years ago in a colossal genesis explosion. It has been expanding at near light speed ever since. In the time it takes to read this sentence, the universe will increase in volume by 100 trillion cubic light-years.

The expansion separates galaxy clusters from one another, like dots on an inflating balloon. This process has made it possible for astronomers to probe into the past, all the way back to the creation era. The expansion has pushed us so far from some sectors of the universe that it takes light from remote galaxies vast amounts of time to reach Earth. These galaxies are seen as they were, not as they are today. Gazing vast distances into space is therefore like entering a cosmic time machine, an enormously useful tool with which to study the evolution of the universe.

The Milky Way Galaxy is thought to be about 13 billion years old. When we look at objects between 10 and 13 billion light-years away, we see them as they were soon after their formation (assuming that all galaxies were born at roughly the same time). These youthful galaxies are far more energetic than our sedate star city, pumping out up to 10,000 times the radiation emitted by the Milky Way Galaxy. No matter where astronomers look at these enormous distances, vigorously energetic objects are seen. They assume that they are witnessing the earliest — and apparently the most violent — stages of galaxy formation a few billion years after the creation of the universe.

When telescopes probe more than 15 billion light-years, they record nothing but a dull energy haze, called the microwave background. This is believed to be the remnant "glow" of the creation explosion. Thus astronomers conclude that we inhabit a universe whose observational extent is defined by its age. The edge of the universe is an edge in time; it is not possible to look back past the beginning.

Exploration of these distant realms billions of light-years away is beyond the capability of amateur equipment. Only the giant instruments of the largest observatories can be used to seek the secrets at the edge of space and time. However, only this final "total universe" box and the supercluster cube before it are off limits to backyard skywatchers. Even with the unaided eye, it is possible to see the Andromeda Galaxy, two million light-years into Cube 9. A small telescope can reveal the brightest galaxies throughout Cube 9 and even a few that are several million light-years into the inner part of Cube 10, but the hunter must know where to look.

That is what much of this book is about — a balanced treatment of the two key elements of amateur astronomy: first, how to locate and identify celestial objects; second, what recent research has revealed about these bodies. This is more than a handbook of the constellations. It is a guide to exploring the universe.

FOCUS ON THE MILKY WAY GALAXY

The Milky Way Galaxy is our hometown in the universe. By cosmic standards, it is a major metropolis, with close to a trillion stellar citizens. The disc-shaped galaxy, about 100,000 light-years in diameter, has a brilliant nuclear bulge, roughly 13,000 light-years thick and 20,000 light-years wide, containing at least 100 to 200 billion stars. In the nuclear region, 10 to 1,000 stars occupy the same volume of space that our sun does in the spiral arms. Stars very close to the nucleus

From a distance of two million light-years, our galaxy would resemble the Andromeda Galaxy, **above**. *The view toward the centre of the Milky Way Galaxy,* **right**, *reveals an impenetrable wall of stars, dust and gas in the Sagittarius Arm, the next spiral arm inward from where the sun resides.*

are less than a quarter of a light-year apart; near collisions must be common, whereas the sun has probably never come within a quarter of a light-year of another star in at least a billion years.

The galaxy's system of spiral arms emerges from the nuclear bulge into a flat, symmetrical pinwheel. The sun is located about two-thirds of the way out from the galaxy's centre, on the inside edge of one of the spiral arms, named the Orion Arm. This is the galaxy's suburban area, well away from the denser nucleus but not too far away from stellar neighbours. The most easily visible of those neighbours are the blue giants and supergiants, such as the stars in Orion's belt. These enormously powerful suns define the spiral arms and give them their bluish white hue.

The yellowish tone to the galaxy's nuclear regions is largely due to the preponderance of yellow and red giants and to the lack of blue giant stars, which are massive suns that squander their stellar fuel at prodigious rates. Blue giants last only a few million years, a lifetime of blazing glory, then evolve to red giants and, soon after, leave the scene, possibly in the blast of a supernova.

Blue giants, and presumably lesser stars like the sun, are born in nebulas, dark clouds of gas and dust woven throughout the spiral arms. Regions where stars are actually forming, such as the Orion Nebula, are visible as bright nebulas, reddish in colour. Several of those visible in binoculars are labelled in the close-up of our quadrant of the Milky Way. That illustration also identifies prominent stars and star clusters, most visible to the unaided eye.

Star clusters are the next evolutionary stage after the bright nebulas. As time passes, the clusters usually disperse, the individual stars spreading throughout the spiral arms. There are nearly as many stars between the spiral arms as in them. But the inter-arm stars are almost all less luminous types, similar to or fainter than the sun. The spiral arms appear to be a shock front or density wave, wherein star formation occurs, the wave somehow propagating throughout the galactic disc. The blue giants and supergiants do not live long enough to drift into the inter-arm zones.

As the sun and other stars in the spiral-arm region orbit the nucleus of the galaxy, they alternately pass through spiral arms and the inter-arm regions. Our sun and planetary system seem to be just entering the Orion Arm. Many of the stars in our immediate vicinity are members of a pocket of fairly young stars called Gould's Belt, on the inside of the Orion Arm. When we look beyond Gould's Belt to the stars in the constellation Orion, we are looking down the Orion Arm toward its trailing edge. When we sight in the direction of the constellation Perseus (the constellations are identified in Chapter 4), we are looking away from the nucleus toward the Perseus Arm. The constellation Cygnus is located in the inward segment of the Orion Arm. In late summer, the Milky Way near the southern horizon is the Sagittarius Arm, the galaxy's centre. Dark nebulas obscure the nuclear bulge and limit our view much beyond the Sagittarius Arm.

The spiral arms of the Milky Way Galaxy contain 100 to 200 billion solar masses of material, 10 to 15 percent of it in the form of nebular dust and gas. Beyond the spiral arms is the galactic halo, the abode of the globular clusters — dense, spherical swarms of up to four million suns. The globulars have huge looping orbits around the galaxy's nucleus. Recent research indicates that the halo — present around other galaxies as well — contains the equivalent mass of hundreds of billions of stars. But few stars are seen. This "invisible" mass is one of modern astronomy's major mysteries.

The structure of the Milky Way Galaxy, as illustrated in the renderings on the following two pages, is based on the most up-to-date findings. But since a direct visual view of the nucleus is blocked by nebular material, other methods, primarily radio-telescope analyses, have filled in the details. Even so, our best picture of the overall structure of our own galaxy is not nearly as clear as routine observatory photographs of other galaxies.

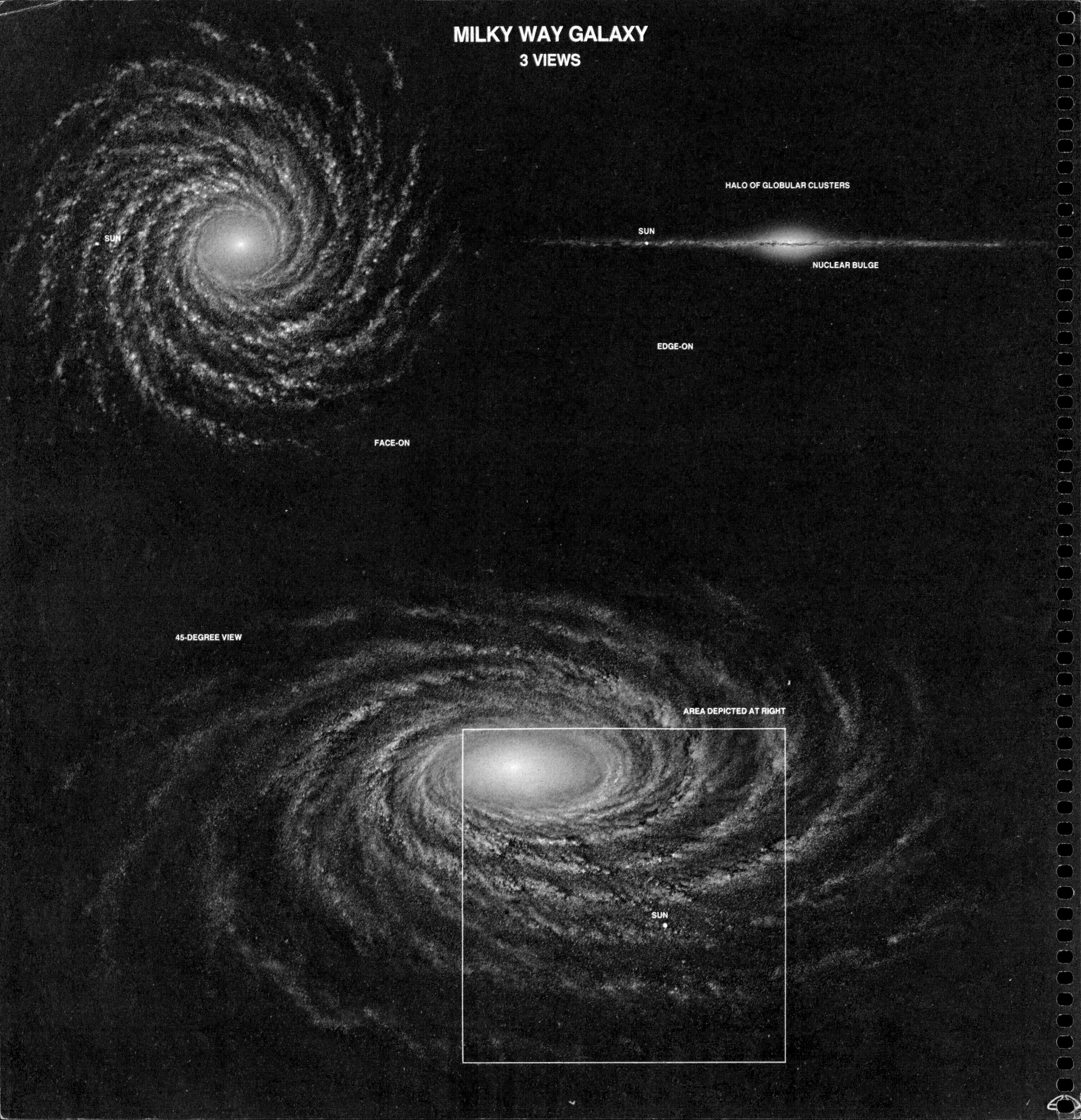
MILKY WAY GALAXY
3 VIEWS
SUN
FACE-ON
HALO OF GLOBULAR CLUSTERS
SUN
NUCLEAR BULGE
EDGE-ON
45-DEGREE VIEW
AREA DEPICTED AT RIGHT
SUN

MILKY WAY GALAXY
NUCLEUS
NUCLEAR
DISC
2 KPC EXPANDING RING
3 KPC ARM
NORMA INTERNAL ARM
CENTAURUS ARM
TR24
"JEWEL BOX"
ETA CARINAE
NEBULA
EAGLE
NEBULA
OMEGA
NEBULA
NGC 6231
SAGITTARIUS ARM
TRIFID
NEBULA
LAGOON
NEBULA
M25
M6
GUM
NEBULA
M7
SUN
CYGNUS LOOP
PRAESEPE
ORION ARM
DENEB
PLEIADES
BELT STARS
M46
RIGEL
ORION
NEBULA
NORTH
AMERICA
NEBULA
ROSETTE
NEBULA
CRAB NEBULA
PERSEUS ARM
PERSEUS
DOUBLE CLUSTER
PERSEUS EXTERNAL ARM

3

Backyard Astronomy

We had the sky up there, all speckled with stars, and we used to lay on our backs and look up at them and discuss about whether they was made or only just happened.

— Mark Twain
Huckleberry Finn

More than a century ago, Ralph Waldo Emerson wrote: "The man in the street does not know a star in the sky." No surveys have ever been conducted to determine by what percentage Emerson was right, but Shakespeare pointed out part of the problem in *Julius Caesar* when he described the night sky as "painted with unnumbered sparks."

Despite its magnificence, the starry night sky can be a bewildering chaos of luminous points. It takes time to sort it out. For every curious stargazer who has learned to distinguish one star from another, there are dozens who have taken a star chart outside and, after an hour or two, have given up in frustration.

The problem is usually the charts, rather than the observer. Although not in use much today, totally impractical charts, such as that shown on the facing page, are still around. Many modern star charts and so-called star finders are either too small or too crowded with dots and lines to be useful during the crucial first nights of star identification.

I vividly remember the thrill of recognizing my first constellation on a clear winter night in 1958. (Constellations are patterns of stars partitioned and named by our ancestors.) I had seen constellation diagrams in astronomy books, but I was intimidated by the complex-looking charts. Then, as I gazed into that crisp winter sky, one image finally crystallized. Standing there before me was the mighty celestial hunter, Orion, his glittering three-star belt framed by four prominent stars marking the shoulders and legs of the legendary nimrod. Every winter since then, Orion has been a familiar friend during the five months or so that he strides across the evening sky.

Orion is the second most prominent stellar configuration in the night sky. Most easily recognized is the Big Dipper, whose seven stars are a familiar sight to many night-sky observers who are unable to identify a single constellation. By using these two star groups, it is possible to identify every major star and constellation visible from Canada, the United States or Europe. Once Orion and the Big Dipper become familiar sights, the rest of the starry sky falls into place, no matter what the season or time of night.

Like our sun, other stars produce their own light and energy through thermonuclear reactions — the same process that generates the fury of a hydrogen bomb. These stars appear as points of light because of their enormous distances, the nearest being about 250,000 times farther away than our sun.

Ninety-nine percent of the stars visible to the unaided eye are bigger and brighter than our sun. This gives a distorted picture of the true variety of suns in space, because the majority of stars in the Milky Way Galaxy are actually less luminous than the sun and are invisible to the backyard stargazer. Those seen with the unaided eye are the titans of the galaxy — searchlights among throngs of 100-watt light bulbs.

The remoteness of these celestial beacons has another important consequence: the stars do not appear to move relative

LEO
Cor Leonis
Cauda Leonis
Jordanis flu.
Tropicus
Mogol
M A R
D
I N D I A
Cor Hydrae Alphard
Crater Vas
140
170
20
10

The backyard astronomer is presented with the illusion of viewing the universe from a flat observing platform, **right**, *but that platform is, of course, attached to the spinning Earth. As Earth rotates in one direction, the sky seems to move the opposite way. Stars in the north pivot counterclockwise around the north celestial pole. Time-exposure photograph,* **left**, *shows this motion as arcing star trails. The axis of the Earth's rotation, when projected into space, passes very close to Polaris, the North Star. Polaris is the small bright arc near centre. The rest of the sky sweeps in a general east-to-west pattern, as the time exposure of the southeastern night sky,* **above**, *reveals. Southern-hemisphere observers see the same east-to-west motion of the dome of stars, but the south polar region swings clockwise.*

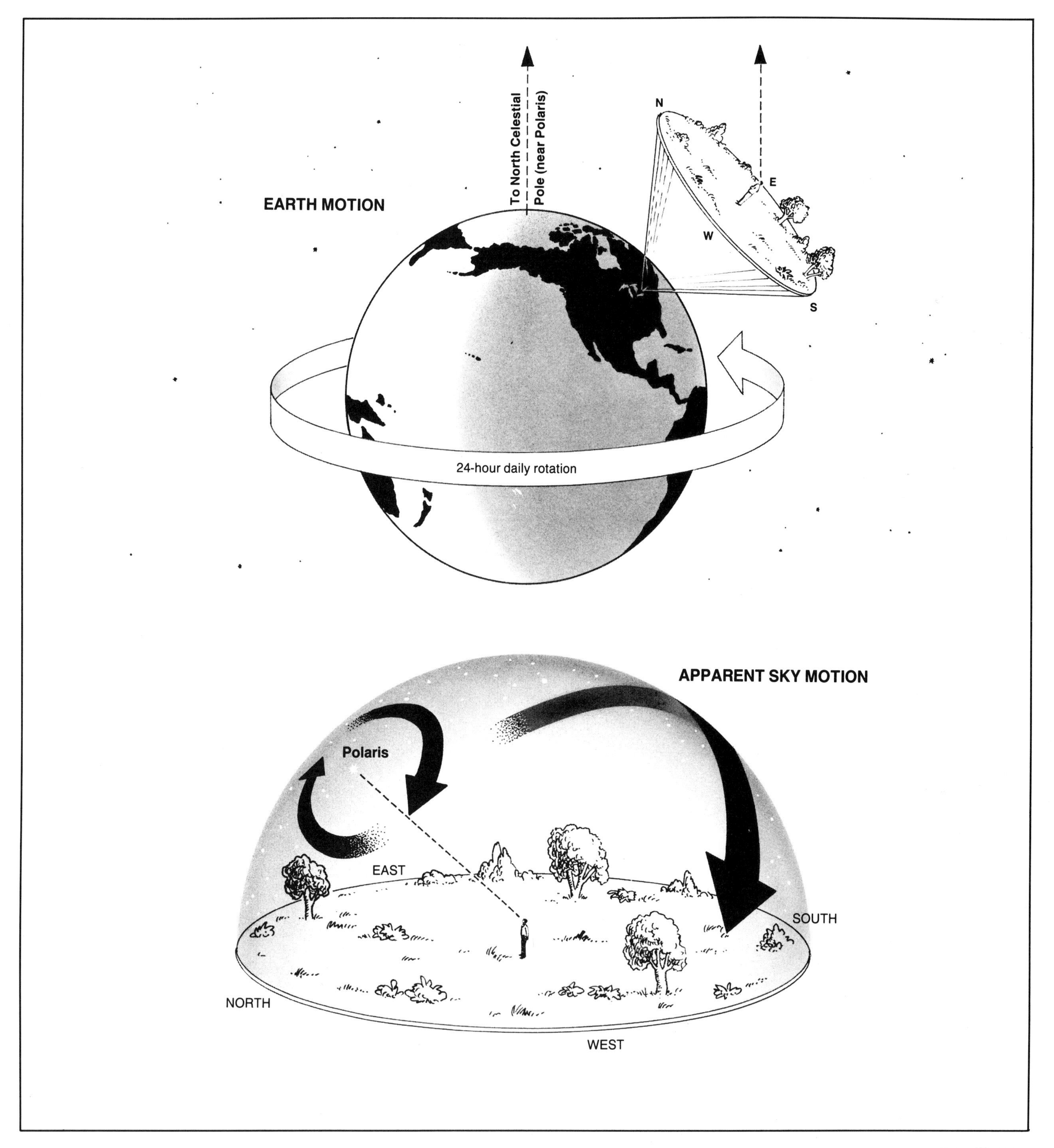
EARTH MOTION
To North Celestial Pole (near Polaris)
N
E
W
S
24-hour daily rotation
APPARENT SKY MOTION
Polaris
EAST
SOUTH
NORTH
WEST

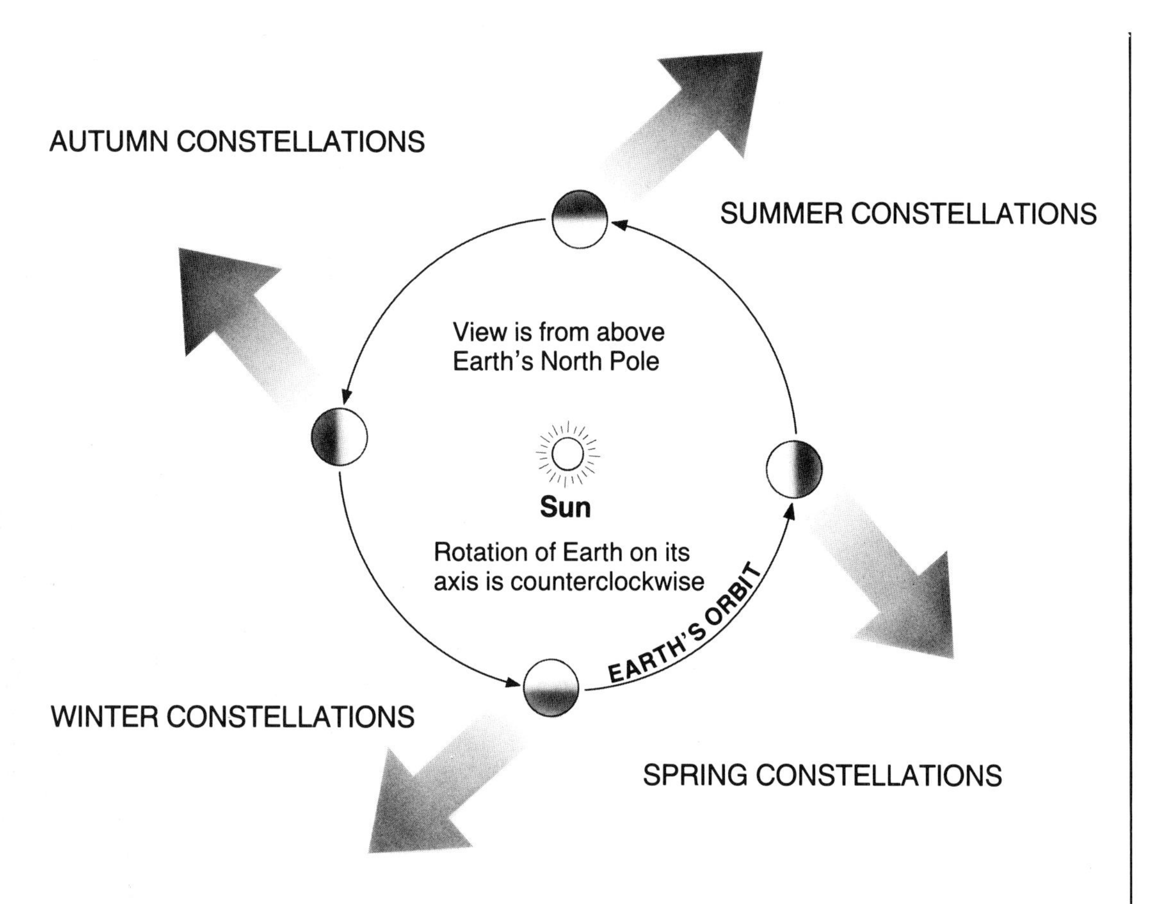

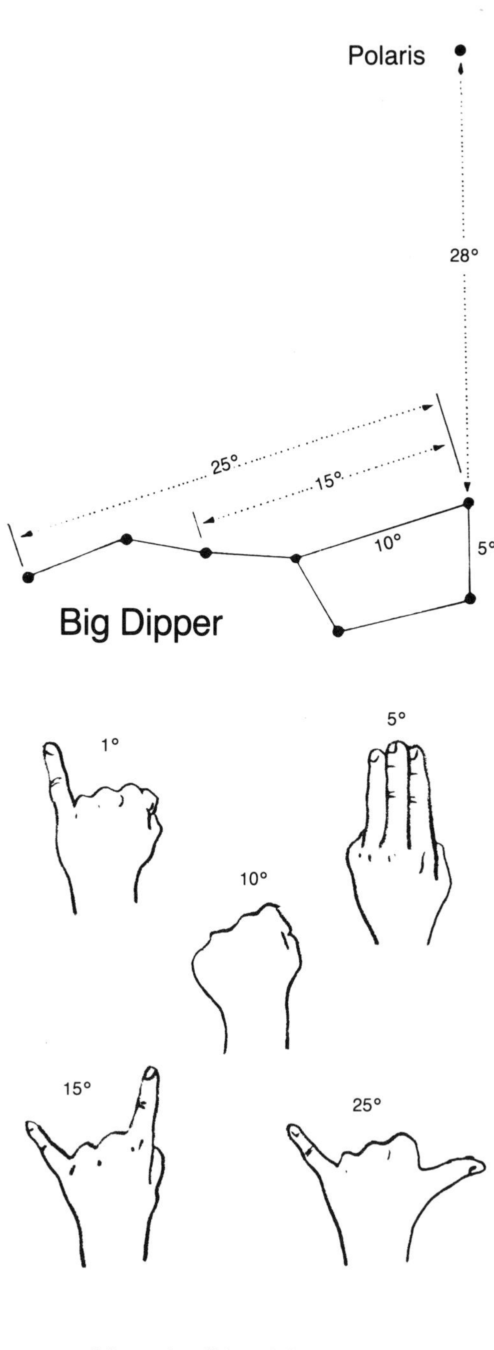

Handy Sky Measures

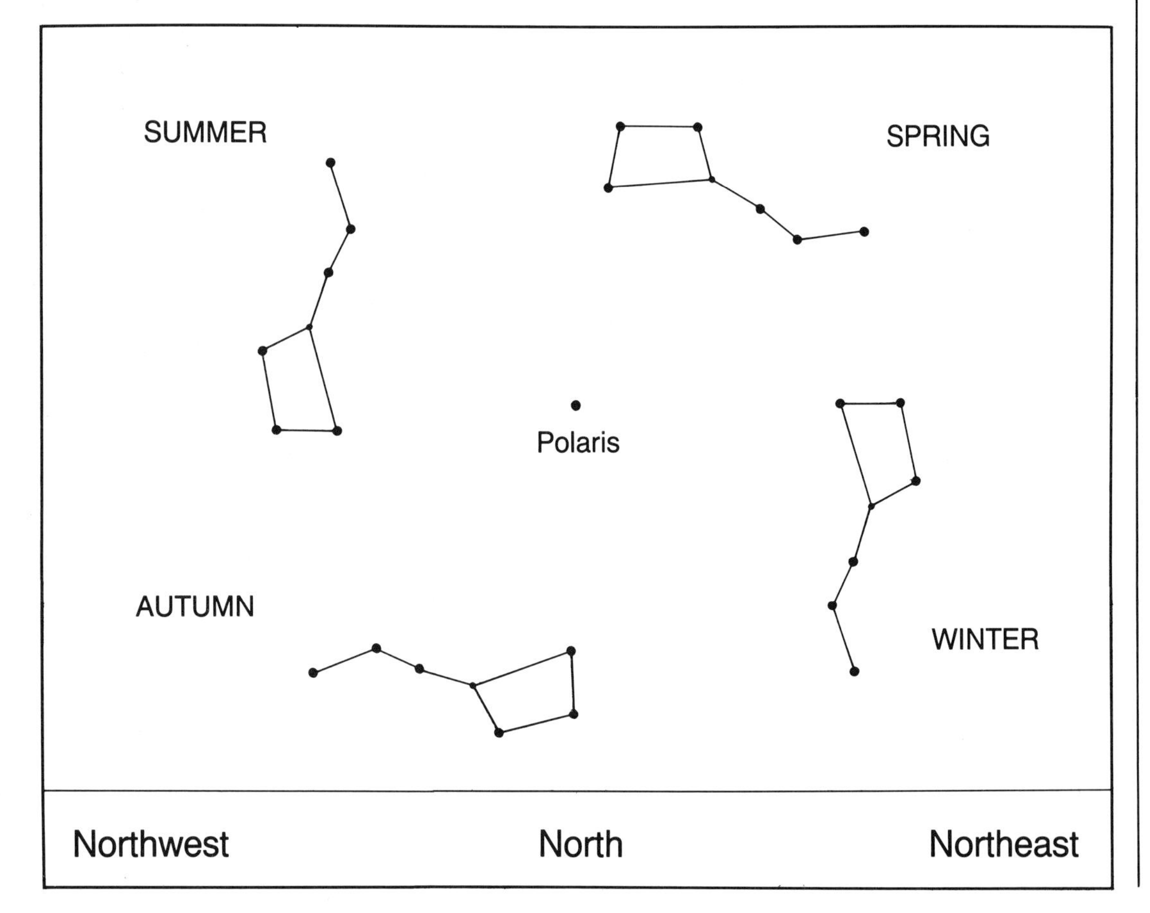

The Big Dipper shifts orientation during the year, **left**, *from close to the northern horizon in autumn to nearly overhead in spring. These major seasonal changes are due to the Earth's annual orbit around the sun,* **above left**, *which points us toward different quadrants of the cosmos as the seasons pass. This action causes many star groups to be well seen only during one season, but the Dipper's northerly location keeps it pivoting the celestial pole and in view all year. As explained in the text, sky distances in degrees can be roughly measured using your hand held at arm's length,* **above**.

to one another from week to week or even from one year to the next. Although they are travelling in space, that movement is insignificant compared with the distances between them. Our great-grandparents saw the Big Dipper stars exactly as they are seen today. Star charts made by the Greek astronomer Hipparchus more than 2,000 years ago show the same stars in almost exactly the same positions as we now see them. (I say "almost" because one bright star and a handful of fainter ones have shifted about the width of the full moon since then. But most have not altered their positions relative to each other by more than the width of a human hair held at arm's length.)

Although their positioning is random, by some fortuitous cosmic coincidence, the stars in the Big Dipper are arranged in a pattern that provides at least seven sky pointers which guide the eye naturally to about a dozen bright stars or stellar groups. The Big Dipper's prime importance as a key to the night sky is its constant availability for midnorthern-hemisphere observing. No matter what day of the year or what time of night, the Big Dipper is seen in the northern part of the sky every clear night of the year from everywhere in Canada and from north of 40 degrees latitude in the United States. (Between 40 and 25 degrees north latitude, the Big Dipper is near or below the northern horizon for several weeks in autumn.) Orion in the southern sector of the sky augments the Big Dipper from November to April.

To locate these guiding constellations, consult the table on page 30. Although the Big Dipper and Orion move around the sky in a systematic way, so does everything else. The only additional information necessary before making the first series of star identifications is how to measure sky distances.

SKY MEASURES

Just as road maps have distance indicators between cities, our celestial guide maps denote distances between key stars and star groups – not distances from Earth to the stars but, rather, the *apparent* distance from one star to another. This measure is calibrated in degrees, and the system is relatively simple. It requires a standard measuring device marked off in multiples of degrees – the human hand. At arm's length, the width of the end of the little finger is almost exactly one degree – wide enough to cover the sun or the moon, both about half a degree across.

The Big Dipper, to the right in this moonlit winter scene, is the sky's best-known stellar landmark. Use the two stars in the Dipper's bowl that are opposite the handle to point to Polaris.

The two pointer stars in the bowl of the Big Dipper, used to find Polaris, are five degrees apart, the width of three fingers held boy scout fashion at arm's length.

For larger sky angles, one fist width is 10 degrees, while 15 degrees is the span between the first and little fingers spread out. An entire hand span, from thumb to

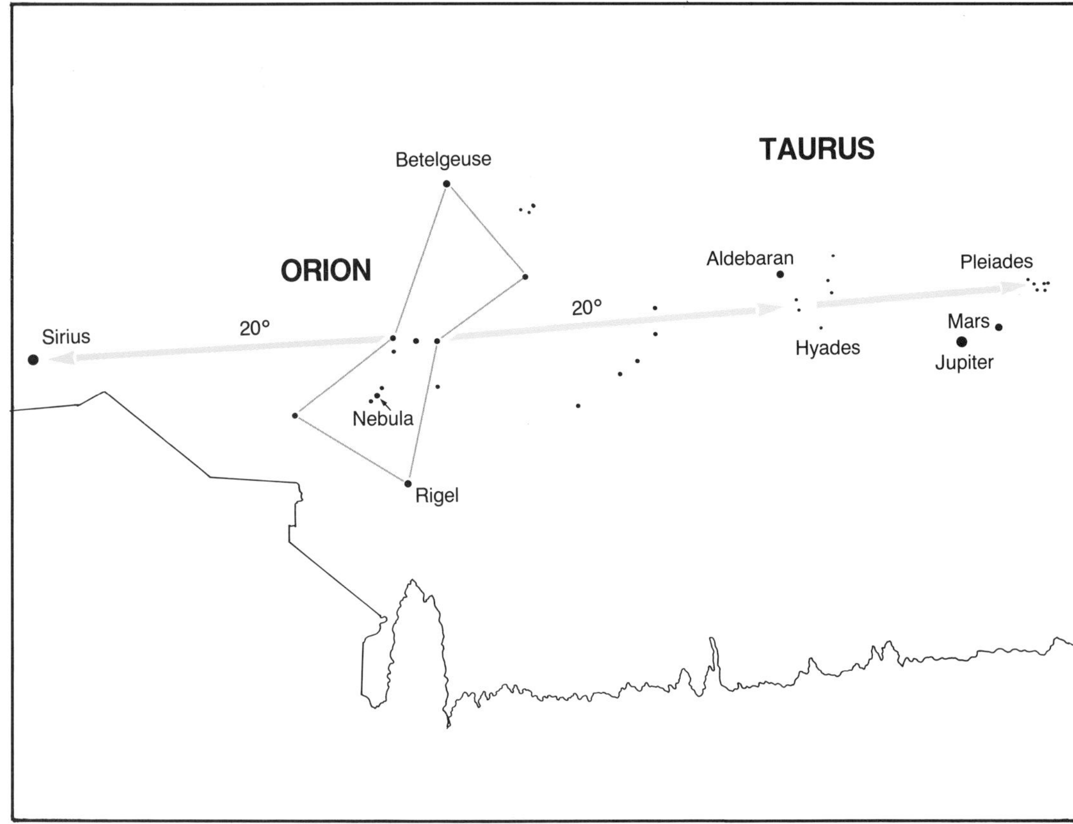

little finger, is about 25 degrees, the length of the Big Dipper. Larger dimensions can be measured in multiples of these. For general reference, the distance from the horizon to overhead is 90 degrees. Remember that these hand-reference measurements work only at arm's length. The system is reasonably accurate for all men, women and children, since people with small hands tend to have shorter arms. Only the hand-span measure seems to vary from person to person because some people can extend their thumb and little finger much more widely than others. A quick check against the Big Dipper will indicate whether you have a span closer to 20 degrees rather than 25. Anyone can become proficient at gauging the distances in degrees from one star or star group to another in a matter of minutes.

It does not matter which season you begin; one or two of the Big Dipper diagrams on page 29 can be used to locate several prominent stars almost instantly once you have a sense of the dimensions involved. This is the crucial first step toward becoming a backyard astronomer. Orion's seven brightest stars — three in the belt and four in a surrounding quadrilateral — are equally efficient as celestial guideposts. Orion's only drawback, compared with the Big Dipper, is that it is prominent in the evening sky only from late November to early April.

Backyard astronomy does not have to be a maze of formulas, calculators, grid lines, nomenclature, mythology and jargon. It can be easy and fun. The all-sky charts in the next chapter are simply extensions of the same principle of using distinctive stellar guideposts to lead the observer around the sky. This is a gradual, painless way to come to know the night sky. Meanwhile, here are a few helpful items of background.

STAR BRIGHTNESS

A glance at the Big Dipper reveals that its stars are not the brightest in the sky. There are several stars substantially brighter and many significantly fainter.

While pondering the starry sky and trying to develop the first star catalogue, the Greek astronomer Hipparchus hit upon the idea of classifying the stars by their brightness. He decided to divide them into six categories. He designated the brightest first magnitude and the faintest sixth, with the others scattered in between.

This system is still in use today, although it has been refined and expanded to include telescopic stars fainter than sixth magnitude and objects brighter than first. A step of one magnitude is an increase or decrease by a factor of 2½ times in brightness. Therefore, an average first-magnitude star is 2½ times brighter than an average second-magnitude star, 6 times brighter than third, 16 times brighter than fourth, 40 times brighter than fifth and 100 times as brilliant as a sixth-magnitude star, the faintest that can be seen on a very clear night with the unaided eye.

No system is perfect. Some stars designated by Hipparchus as first magnitude are too bright. They are now rated zero, those brighter still are −1, and so on. (The scale is like an upside-down thermometer.) The magnitude classes extend in the faint direction all the way to 28th magnitude, the dimmest objects that can be detected with the world's largest telescopes. A sixth-magnitude star is 15 million times brighter than one of 24th magnitude. At the opposite extreme, the brightness of the sun is rated at magnitude −26, six trillion times brighter than a sixth-magnitude star.

Sirius, the brightest star in the night sky, is magnitude −1. Only Jupiter (−3), Venus (−4) and Mars (varies from +2 to −3) are brighter.

CONSTELLATION & STAR NAMES

Long before Hipparchus, skywatchers of antiquity divided the sky into groups of stars called constellations. The stars in a constellation are seldom related to one another. These celestial groupings are steeped in mythology and, in the case of the zodiac constellations, embroidered with the symbolism of astrology.

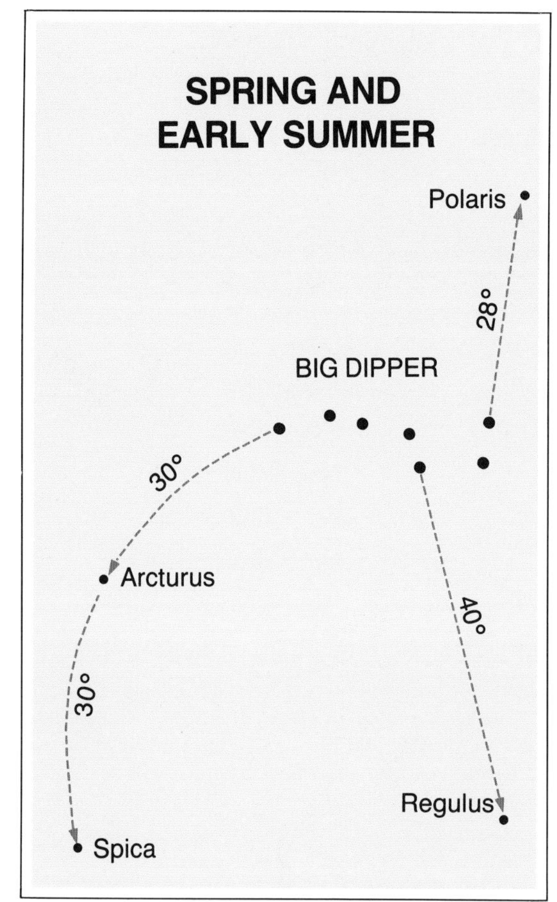

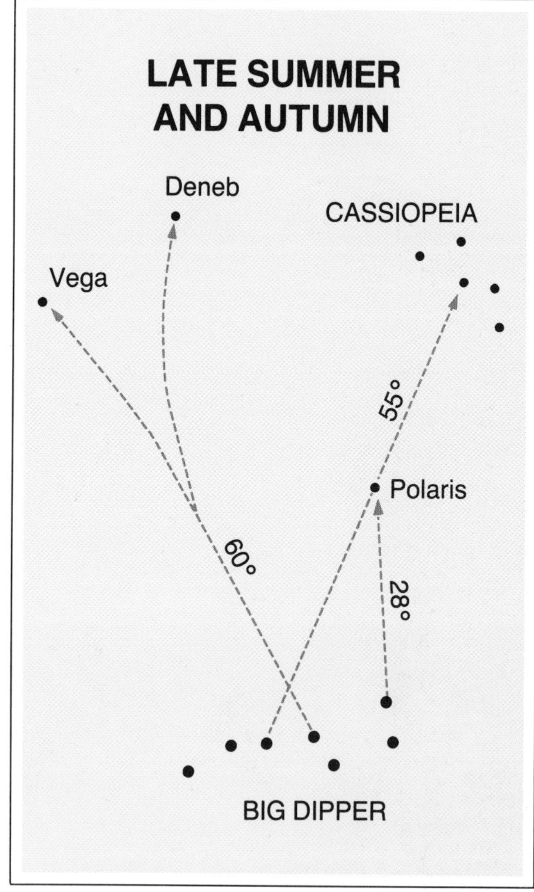

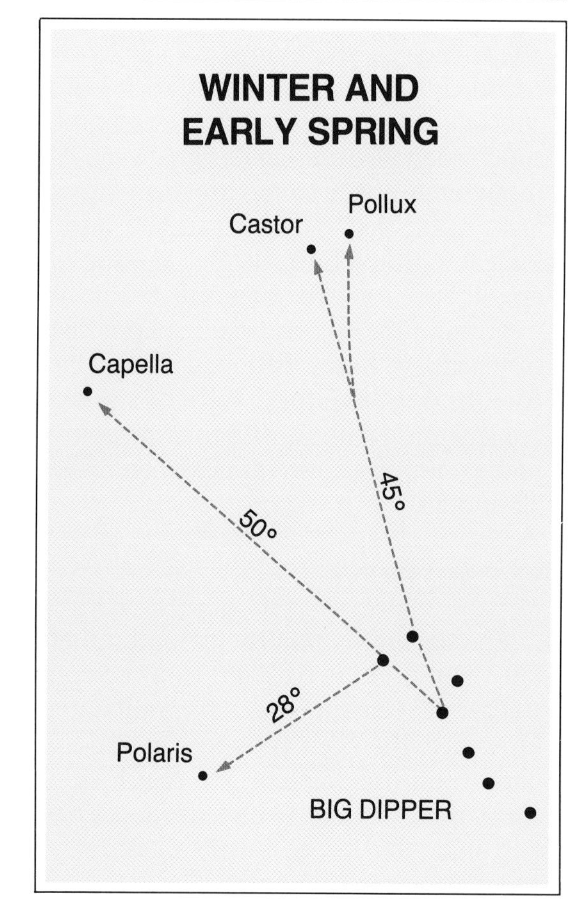

The crucial first step in locating stars and constellations is to identify the Big Dipper and Orion and then use them as guides, as shown on these two pages. Orion is visible in the evening sky from late November until early April. The photograph of Orion, **facing page**, *was taken in March, but the constellation is similarly oriented at other times so that the locater arrows to Sirius and to the Hyades and Pleiades star clusters in Taurus always work. (The planets Jupiter and Mars move among the stars and thus will not be seen in positions shown.) The Big Dipper diagrams,* **this page**, *clearly show why it is the sky's premier guidepost. It is visible every night of the year (except for a brief period in autumn from the southern United States) and serves as an unsurpassed pointer. All stars shown in the three Big Dipper diagrams are as bright as or brighter than the Dipper stars.*

WHERE TO FIND THE BIG DIPPER AND ORION

MONTH	BIG DIPPER		ORION	
	Direction	Altitude	Direction	Altitude
January	NE	25°	S	40°
February	NE	40	S	45
March	NE	55	SW	35
April	N	65	SW	20
May	N	70	Not visible	
June	N	65	Not visible	
July	NW	55	Not visible	
August	NW	40	Not visible	
September	NW	25	Not visible	
October	N	15	Not visible	
November	N	10	E	15
December	N	15	SE	30

Information intended for midevening 8 to 10 p.m., local standard time, or 9 to 11 p.m., local daylight time; latitude 40 to 55 degrees N. In more southerly latitudes, the Big Dipper will appear proportionately lower and Orion higher in the sky.

Today, astronomers still use the traditional Latin forms of the constellation names from classical Greek civilization. A few constellations (mostly dim ones) were invented in the 17th and 18th centuries to fill in sky regions not included in the ancient lore. In 1930, the constellation names and boundaries were officially set by the International Astronomical Union, and there have been no changes since. Although there are 88 constellations, a quarter of them are in the southern sky, concealed from view in midnorthern latitudes; half of the remainder are rather faint. Being able to identify the dimmer constellations is not necessary in the beginning stages of amateur astronomy.

The situation with individual star names is less formal. Several hundred of the brightest stars have been named over the centuries, but only about 75 of these designations have survived the dustbin of disuse. As far as we know, the Babylonians were the first to name stars, but most of the names used today are Arabic, with a sprinkling of Greek, Latin and Persian. The preponderance of Arabic star names is due to the fact that during the Dark Ages, Arabic astronomy was the most advanced in the world. The Arabs retained the tradition of Greek-Latin constellation names, but their star names superseded most earlier designations. Although many star names are meaningless in English, they usually translate into a logical word picture. Betelgeuse, for example, is believed to be ancient Arabic for "armpit of the mighty one." Spica is "ear of wheat" in Latin; Procyon means "before the dog" in Greek. A list of star and constellation names, with their meanings and pronunciations, appears on pages 32 and 33.

SKY MOTIONS

The stars move relative to one another, but the motion is so slight that the constellations remain intact for millennia. However, the sky as a whole has an apparent motion caused by the Earth's spinning on its axis (which produces day and night) and by the Earth's annual orbit around the sun (our year). The daily axis rotation means that at different times of the night, any one point on Earth faces in different directions. You can observe this motion on a clear night by standing so that a bright star is just above a marker object – a pole or house gable, for example. Note the position, and return to that spot 15 to 45 minutes later. The star will have moved by a noticeable amount. In general, the stars swing from east to west, just as the sun does each day. Stars in the north slowly twirl around the sky's pivot point, the north celestial pole, not far from Polaris, which hardly moves at all. The effect is vividly illustrated in the time-exposure photographs in this chapter.

The change of stellar scenery caused by the Earth's orbital trek around the sun becomes apparent only over weeks or months, but the consequences are more profound. Whole sectors of the sky are paraded into view, changing just as the seasons do. Each season, therefore, has its prominent stars and constellations, which are illustrated in the following chapter.

Throughout this book, and in general astronomy usage, the stars and constellations referred to in each season are those that the *evening* side of Earth is pointed toward. Since half the entire sky is seen at any one time (Earth itself blocks the other half), some overlap occurs. In the spring sky, for example, a few winter and summer stars are also seen. By staying up all night, an observer gets a jump on the seasons, at least in a celestial context. As Earth rotates on its axis (counterclockwise, viewed from the north side, the same direction as its orbital revolution), the stars of the following season begin to appear in the east starting around midnight. By 4 a.m., the observer has been transported one whole season ahead – that is, Earth has rotated to face the part of the sky seen in the evening hours of the following season.

Within two minutes, three Earth-orbiting satellites cruised across a small sector of the constellation Hercules. A sharp-eyed skywatcher can see more than 10 satellites during the first hour after darkness falls.

ERSATZ STARS

There is more to a starry night than just stars, planets and the moon. Pulsating red and white aircraft lights occasionally punctuate the scene. Darting flashes signal the incineration of meteors, bits of cosmic debris entering the Earth's atmosphere. And there are numerous moving lights, slower than most aircraft, usually steady in brightness and tracking in apparently slightly wavy trajectories.

These moving beacons can travel in any direction, their unblinking white light deceivingly like a moderately bright star, except for their motion. At least one of these objects can normally be spotted after one scans the skies for 15 minutes on a dark night, especially in the hour after dusk, when the stars are first visible. They are artificial satellites in Earth orbit. Those seen with the unaided eye range from the size of a refrigerator to that of a good-sized house trailer.

When the U.S. Skylab space station plummeted to Earth over Australia in 1979, our brightest artificial satellite disappeared. Sunlight reflecting from the 75-ton space station occasionally produced a "star" rivalling Vega or Arcturus. Today, the Soviet Mir space station is almost as bright, and hundreds of others are visible to the unaided eye.

In the early 1960s, the United States launched two colossal Mylar balloons into orbit. These celestial baubles were coated with an ultrathin, highly reflective aluminum skin. Echo 2, the larger of the pair, was like a roaming Jupiter coursing through the constellations. The Echo satellites were used to reflect communications signals between Earth stations. They collapsed after a few years in orbit and, in any event, were superseded by modern communications satellites.

Most communications satellites are lined up above the Earth's equator in 24,000-mile-high orbits, which match our planet's rotation; thus each seems to hover over one location. At such distances, they are invisible to the unaided eye. Spy satellites are different. Since the military wants to get as close to Earth as possible, these satellites skim just beyond the fringe of the atmosphere, about 120 miles up. They move across the sky in about five minutes, reflecting sunlight off their metallic bodies.

That is why most satellites are seen just as it gets dark; there is still plenty of sunlight flooding the few hundred miles overhead. If the vehicle is tumbling, indicating that it is either out of action or a discarded section of a launching rocket, it will pulse in a regular rhythm.

Whether satellites have a steady or fluctuating brightness, most novice observers agree that they do not appear to move across the sky in perfectly straight lines. There seems to be a perceptible waviness to their paths, a jerkiness in speed as they glide through the starry background. In fact, these oscillations are in the mind, not the sky. The satellites actually move in precise linear paths at an even velocity.

The human brain likes to link patterns into a recognizable image. This is done instantaneously in daily life. However, looking at one moving light in a randomly dotted black sky, the brain constantly tries to produce these patterns but fails. What are thought to be oscillations in the satellite's path are really the unconscious workings of the mind trying to make sense out of an unfamiliar visual environment. The result is, in effect, an optical illusion.

Star and Constellation Pronunciation Guide

Name	Object	Pronunciation	Meaning*
Acamar	star in Eridanus	AKE-uh-mar	end of river (A)
Achernar	brightest star in Eridanus	AKE-er-nar	end of river (A)
Adhara	star in Canis Major	ah-DARE-rah	the maiden (A)
Albireo	star in Cygnus	al-BEER-ee-oh	[meaning unknown]
Alcor	star in Ursa Major	AL-core	the abject one (P)
Alcyone	brightest star in Pleiades	al-SIGH-oh-nee	one of the Pleiades sisters from mythology (G)
Aldebaran	brightest star in Taurus	al-DEB-uh-ran	the follower [of the Pleiades] (A)
Alderamin	star in Cepheus	al-DARE-uh-min	the right forearm (A)
Algeiba	star in Leo	al-JEE-buh	the forehead (A)
Algenib	star in Pegasus	al-JEE-nib	the flank (A)
Algol	variable star in Perseus	AL-gall	the ghoul (A)
Alhena	star in Gemini	al-HE-na	the brand mark (A)
Alioth	star in Big Dipper	ALLEY-oth	the goat (A)
Alkaid	star in Big Dipper	al-KADE	daughter of the bear (A)
Almach	star in Andromeda	AL-mac	the weasel (A)
Alnair	brightest star in Grus	al-NAIR	the bright (A)
Alnilam	star in Orion's belt	al-NIGH-lam	the arrangement [of pearls] (A)
Alnitak	star in Orion's belt	al-NIGH-tak	the belt (A)
Alpha Centauri	brightest star in Centaurus	AL-fah sen-TORE-eye	[modern designation]
Alphard	brightest star in Hydra	AL-fard	the solitary (A)
Alphecca	brightest star in Corona Borealis	al-FECK-ah	the broken ring [of stars] (A)
Alpheratz	star in Andromeda	al-FEE-rats	navel of the steed (A)
Altair	brightest star in Aquila	al-TAIR	the flying one (A)
Andromeda	prominent constellation	an-DROM-eh-duh	daughter of Cassiopeia in mythology (G)
Antares	brightest star in Scorpius	an-TAIR-eez	rival of Mars (A)
Aquarius	zodiac constellation	a-QUAIR-ee-us	the water carrier (L)
Aquila	prominent constellation	A-quill-ah	the eagle (L)
Arcturus	brightest star in Bootes	ark-TOUR-us	bear guard (G)
Aries	zodiac constellation	AH-rih-eez (or AIR-eez)	the ram (L)
Arneb	brightest star in Lepus	AR-neb	the hare (A)
Auriga	prominent constellation	oh-RYE-gah	the charioteer (L)
Bellatrix	star in Orion	bell-LAY-trix	the warrioress (L)
Betelgeuse	star in Orion	BET-el-jews	armpit of the mighty one (A)
Bootes	prominent constellation	bo-OH-teez	the herdsman (G)
Canes Venatici	small constellation	KAY-neez vee-NAT-ih-sigh	the hunting dogs (L)
Canis Major	prominent constellation	KAY-niss MAY-jer	the great dog (L)
Canis Minor	small constellation	KAY-niss MY-ner	the lesser dog (L)
Canopus	brightest star in Carina	can-OH-pus	the helmsman (G)
Capella	brightest star in Auriga	kah-PELL-ah	she-goat (L)
Caph	star in Cassiopeia	kaf	the hand (A)
Carina	prominent southern constellation	ka-RYE-nah (or ka-REE-nah)	the keel [of the ship Argo] (L)
Cassiopeia	prominent constellation	KAS-ee-oh-PEE-ah	wife of Cepheus in mythology (G)
Castor	star in Gemini	KAS-ter	beaver (G)
Centaurus	prominent southern constellation	sen-TOR-us	the centaur (G)
Cepheus	constellation	SEE-fee-us (or SEE-fuce)	king of Ethiopia in mythology (G)
Cetus	large, dim constellation	SEE-tus	the whale menacing Andromeda (G)
Coma Berenices	small constellation	KOH-mah berry-NICE-eez	Berenice's hair (G)
Cor Caroli	brightest star in Canes Venatici	kor CARE-oh-lie	heart of Charles [Charles II of England] (L)
Corona Borealis	small constellation	kor-OH-nah bo-ree-ALICE	the northern crown (L)
Corvus	small constellation	CORE-vus	the crow (L)
Cygnus	prominent constellation	SIG-nus	the swan (G & L)
Delphinus	small constellation	del-FINE-us	the dolphin (G & L)
Delta Cephei	variable star in Cepheus	DEL-ta SEE-fee-eye	[an important variable star]
Deneb	brightest star in Cygnus	DEN-eb	tail of the hen (A)
Denebola	star in Leo	duh-NEB-oh-lah	tail of the lion (A)
Diphda	brightest star in Cetus	DIFF-dah	the frog (A)
Draco	constellation	DRAY-ko	the dragon (G)
Dschubba	star in Scorpius	JEW-bah	the forehead (A)
Dubhe	star in Big Dipper	DUE-bee	the bear (A)
Elnath	star in Taurus	el-NATH	the butting [horn] (A)
Eltanin	star in Draco	el-TAY-nin	the sea monster (A)
Enif	star in Pegasus	ENN-if	the nose [of the horse] (A)
Equuleus	small constellation	ee-KWOO-lee-us	the little horse (L)

Name	Object	Pronunciation	Meaning*
Eridanus	constellation	eh-RID-uh-nuss	a river (G)
Fomalhaut	brightest star in Piscis Austrinus	FOAM-a-lot (or FOAM-ah-low)	mouth of the fish (A)
Gemini	zodiac constellation	GEM-in-eye (or GEM-in-knee)	the twins (G)
Hadar	star in Centaurus	HAD-ar	the settled land (A)
Hamal	brightest star in Aries	HAM-el	the ram (A)
Hyades	star cluster in Taurus	HI-a-deez	half-sisters to the Pleiades (G)
Izar	star in Bootes	EYES-ar	the loincloth (A)
Kochab	star in Ursa Minor	KOE-kab	the star (A)
Lacerta	small constellation	la-SIR-tah	the lizard (L)
Lepus	constellation	LEE-puss	the hare (L)
Libra	zodiac constellation	LYE-bra (or LEE-bra)	the balance (L)
Lupus	constellation	LEW-puss	the wolf (L)
Lyra	prominent constellation	LYE-rah	the lyre (G)
Markab	star in Pegasus	MAR-keb	the part for riding on (A)
Megrez	star in Big Dipper	ME-grez	the insertion point [of the bear's tail] (A)
Menkalinan	star in Auriga	men-KAL-in-nan	shoulder of the charioteer (A)
Menkar	star in Cetus	MEN-kar	the nostril [of the whale] (A)
Menkent	star in Centaurus	MEN-kent	[modern corruption for Centaurus's shoulder]
Merak	star in Big Dipper	ME-rac	the loins [of the bear] (A)
Mintaka	star in Orion's belt	min-TAK-uh	the belt (A)
Mira	variable star in Cetus	MY-rah	the wonderful (A)
Mirfak	brightest star in Perseus	MUR-fak	the elbow (A)
Mirzam	star in Canis Major	MUR-zam	the roarer [announcing Sirius] (A)
Mizar	star in Ursa Major	MY-zar	the wrapping (A)
Monoceros	constellation	mon-OSS-err-us	the unicorn (G)
Nunki	star in Sagittarius	NUN-key	Sumerian for god of the waters
Ophiuchus	constellation	off-ih-YOU-kus	the serpent bearer (G)
Orion	prominent constellation	oh-RYE-un	the hunter (G)
Pegasus	prominent constellation	PEG-uh-sus	the winged horse (G)
Perseus	prominent constellation	PURR-see-us (or PURR-SOOS)	hero; rescuer of Andromeda (G)
Phact	star in constellation Columba	fact	the dove (A)
Phecda	star in Big Dipper	FECK-dah	the thigh [of the little bear] (A)
Pisces	zodiac constellation	PIE-sees	the [two] fishes (L)
Piscis Austrinus	constellation	PIE-sis OSS-TRY-nus	the southern fish (L)
Pleiades	star cluster in Taurus	PLEE-ah-deez	the seven sisters (G)
Polaris	the North Star	poh-LAIR-iss	[north] pole star (L)
Pollux	brightest star in Gemini	PAW-lux	much wine (L)
Porrima	star in Virgo	poh-RIM-ah	goddess of childbirth (L)
Praesepe	star cluster in Cancer	pray-SEEP-ee	the manger (L)
Procyon	brightest star in Canis Minor	PRO-see-on	before the dog (G)
Rasalgethi	star in Hercules	ras-el-GEE-thee	head of kneeling one (A)
Rasalhague	star in Ophiuchus	RAS-el-HAY-gwee	head of the snake man (A)
Regulus	brightest star in Leo	RAY-gu-lus	prince (L)
Rigel	brightest star in Orion	RYE-jel	the foot (A)
Sabik	star in Ophiuchus	SAY-bik	the preceding (A)
Sadr	star in Cygnus	SAD-er	the breast [of the swan] (A)
Sagitta	small constellation	sah-JIT-tah	the arrow (L)
Sagittarius	prominent zodiac constellation	saj-ih-TAIR-ee-us	the archer (L)
Saiph	star in Orion	saw-EEF (or safe)	the sword (A)
Scheat	star in Pegasus	SHEE-at	the leg (A)
Schedar	star in Cassiopeia	SHED-ar	the breast (A)
Scorpius	prominent zodiac constellation	SKOR-pee-us	the scorpion (G)
Scutum	small constellation	SKEW-tum	the shield (L)
Shaula	star in Scorpius	SHOAL-ah	the raised [tail] (A)
Sirius	brightest star in Canis Major	SEAR-ee-us	scorching (G)
Spica	brightest star in Virgo	SPIKE-ah	ear of wheat [held by Virgo] (L)
Tarazed	star in Aquila	TAR-uh-zed	plundering falcon (P)
Taurus	prominent zodiac constellation	TOR-us	the bull (G)
Thuban	star in Draco	THEW-ban	the snake (A)
Vega	brightest star in Lyra	VAY-gah (or VEE-gah)	the stooping [eagle] (A)
Virgo	prominent zodiac constellation	VURR-go	the maiden (L)
Vulpecula	small constellation	vul-PECK-you-lah	the fox (L)
Zubenelgenubi	star in Libra	ZOO-ben-ell-jen-NEW-bee	the southern claw (A)
Zubeneschamali	star in Libra	ZOO-ben-ess-sha-MAY-lee	the northern claw (A)

**A = Arabic, G = Greek, L = Latin, P = Persian; derivation of meaning is sometimes a very rough translation*

4

Stars for All Seasons

Who were they, what lonely men,
Imposed on the fact of night,
The fiction of the constellations?

— Patric Dickinson

Besides being a pleasant way to pass a mild evening under the night sky, learning to identify the stars and constellations is the foundation for all other elements of backyard exploration of the universe. The nightly canopy of stars is a celestial map with which the observer must be familiar before seeking out specific targets for binocular and telescope viewing. The basic elements of star identification were outlined in the last chapter. Now comes the integration to the complete night sky.

All too often, sky charts intended for beginners sacrifice realism and clarity by adding celestial grids, telescopic objects and constellation and star names. This book utilizes a unique all-sky dual-chart system. Full-colour charts for each season, showing the stars as realistically as possible, are paired with identical charts that include names and the complete locater-arrow system introduced in Chapter 3. When used together, these charts should surmount the problems inherent in many sky charts in the past.

The all-sky colour charts are a miniature planetarium in a book. They duplicate the appearance of the sky seen from a typical dark, but not necessarily pitch-black, location in southern Canada or the United States (except the extreme southern states) and most of Europe. The locater-arrow system has unrestricted utility. It will work anywhere in the northern hemisphere.

Each chart is keyed to the evening hours of a particular season, but on almost any night of the year, two of the charts will be usable — one for the evening hours and one for the early morning. Presenting the entire visible sky in one illustration allows rapid linking of one star group to another, and gradually, the night sky will become woven into a single mental picture.

In essence, the all-sky seasonal charts reproduce the dome of the night sky on a flat surface. Thus the horizon becomes the edge of the chart, and the overhead point is at the centre.

The charts are most practical if used a section at a time. Human eyes are not capable of taking in the entire sky at once anyway. Normally, only about a quarter of the whole celestial dome is comfortably viewed without head movement. In order to make the chart conform to any quadrant of the sky, the book must be turned so that the compass point faced by the observer is down. For example, when one is using the summer chart and facing west, the book should be rotated 90 degrees clockwise so that the west compass point is at the bottom. The curving horizon line then corresponds to the actual horizon and looks like the illustration on page 37. The identities of the stars and constellations in this same region are supplied on the black and white chart below the photograph.

If you move around the horizon a wedge at a time, the big picture should soon come into focus. For each season, there is a preferred starting point, usually involving the Big Dipper or Orion. A step-by-step approach is described in the following pages, but first, let us consider the charts.

All stars down to third magnitude and many of the fourth magnitude are shown on each seasonal all-sky chart. The charts would become a bewildering maze of dots if fifth- and sixth-magnitude stars were included. (However, the full sky to fifth magnitude is detailed in the series of 20 charts in Chapter 6.) Here are some suggestions for getting the most use out of the seasonal all-sky charts:

1. Although the charts are designed for specific time spans, they are still useful up to an hour on either side of the intervals indicated, except for locating objects near the horizon.

2. The book's spiral binding allows the facing all-sky charts to be folded back-to-back for outdoor use. One side is the real-sky simulation, the other provides the identification of the stars and constellations.

3. When just starting out, avoid hazy skies or nights washed by the full moon's light. Too few stars will be visible on these nights for proper identification. Conversely, pitch-black skies, though inspiring, sometimes reveal such a profusion of stars that initial identification is difficult. A site protected from direct house or street lighting and not illuminated by serious urban glow is ideal.

4. Work from the known to the unknown. Start with familiar stars and constellations, and utilize the locater arrows pointing to bright stars and new constellations. Be patient – it usually requires at least a year to become completely comfortable with star identification.

5. If a bright star is seen in the vicinity of the line labelled "ecliptic," it is almost certainly a planet. (The ecliptic is the celestial pathway of the sun, moon and planets.) Sorting out the identities of the five naked-eye planets is described in Chapter 7.

6. When you are using this book outdoors, a flashlight filtered with red or deep orange plastic or cellophane or a flashlight that has an extremely dim bulb is recommended. Turning on a bright light to see the chart will destroy the eyes' sensitivity to the darkness, and it will take several minutes for them to recover.

7. Dim indoor lights before going outside. The eyes will become sensitive to low light levels much more quickly, allowing fainter stars to be seen sooner. While outside, avoid looking at streetlights and house lights as much as possible. Not only does direct artificial illumination spoil the aesthetic appeal of the sky, but such lighting also affects the eyes' sensitivity to the dark. (See page 82.)

THE SPRING SKY

My normal enthusiasm for astronomy always gets an extra boost on that first mild spring evening when the stars shine and conditions bode well for a long season of skygazing. The Big Dipper is nearly overhead throughout the spring evenings, its elaborate system of pointer stars providing the best possible opportunity to link up all of the major stars and star groups above the horizon.

The Big Dipper is actually part of the constellation Ursa Major, the large bear in mythology that guards the polar regions. The complete constellation includes the third- and fourth-magnitude stars to the south and west of the Dipper. The Big Dipper itself is the invention of 19th-century American stargazers. In Britain, it is called the Plough. North American Indians pictured the bowl as a bear and the three handle stars as a trio of braves stalking the beast.

A mental extension of the curve of the Big Dipper's handle one full Dipper length reaches zero-magnitude Arcturus, the brightest star in the spring skies. Arcturus is the most prominent star in the constellation Bootes, the herdsman. Its name and location can be memorized with the phrase "follow the arc to Arcturus," which refers to the arching curve made by extending the Big Dipper's handle. Sometimes added to this is "and speed on to Spica," which can be done easily by extending the curve another Dipper length to the first-magnitude star Spica, in the large zodiac constellation Virgo. And the curve does not end there. Another 15-degree extension leads to a small but conspicuous quadrilateral of third-magnitude stars known as Corvus. Corvus's identity can be confirmed by using its top two stars as pointers back to Spica.

The two stars in the Big Dipper's bowl, nearest the handle, can be used to form a locater arrow running 45 degrees south to Regulus, the first-magnitude star in Leo, the lion, one of the few constellations that look something like what they were named for. A backward question mark signifies the beast's head and mane, while Regulus is Leo's heart. His hindquarters are designated by a triangle of stars to the east. Overall, the stars of Leo cover an area of sky slightly larger than the Big Dipper.

The locater arrow that traces diagonally across the Dipper's bowl to Castor and Pollux in Gemini is an important link between spring and winter constellations. In its annual orbit about the sun, Earth has moved to face the region in space decorated by the stars of Leo, Bootes and Virgo. The winter groups are seen low in the west. By summer, Earth will be facing an area in space exactly opposite Gemini, and all the winter stars will be buried in the solar glare.

Of the three brightest stars in spring skies, Regulus and Spica are first magnitude and Arcturus is zero magnitude. Regulus is a bluish star about 85 light-years distant (although it appears white to the eye), with a luminosity about 160 times the sun's. Spica is actually 10 times as bright as Regulus and 4 times as remote.

Arcturus, only 37 light-years away, is one of the nearest bright stars. It is a giant star about 23 times the diameter of our sun and radiating about 130 times as much energy. Its pale orange colour is evident even to the unaided eye.

The Pleiades, **previous page**, *is a cluster of young stars 410 light-years away that is visible to the unaided eye in autumn skies.* **Right:** *The circular all-sky charts are best utilized a section at a time, as shown here, to correspond to the direction the observer is facing.*

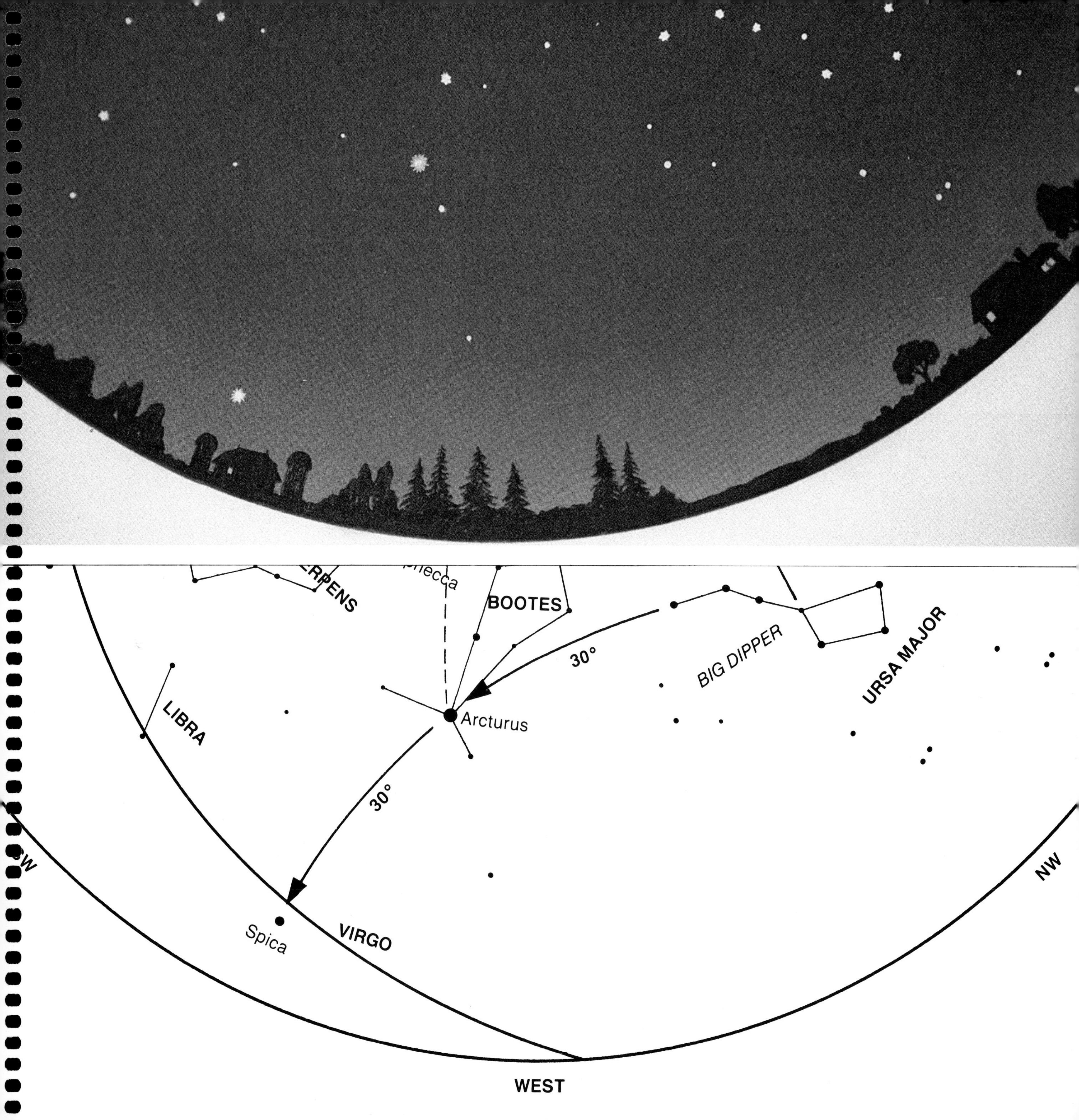
BOOTES
BIG DIPPER
URSA MAJOR
30°
LIBRA
Arcturus
30°
NW
Spica
VIRGO
WEST

Spring

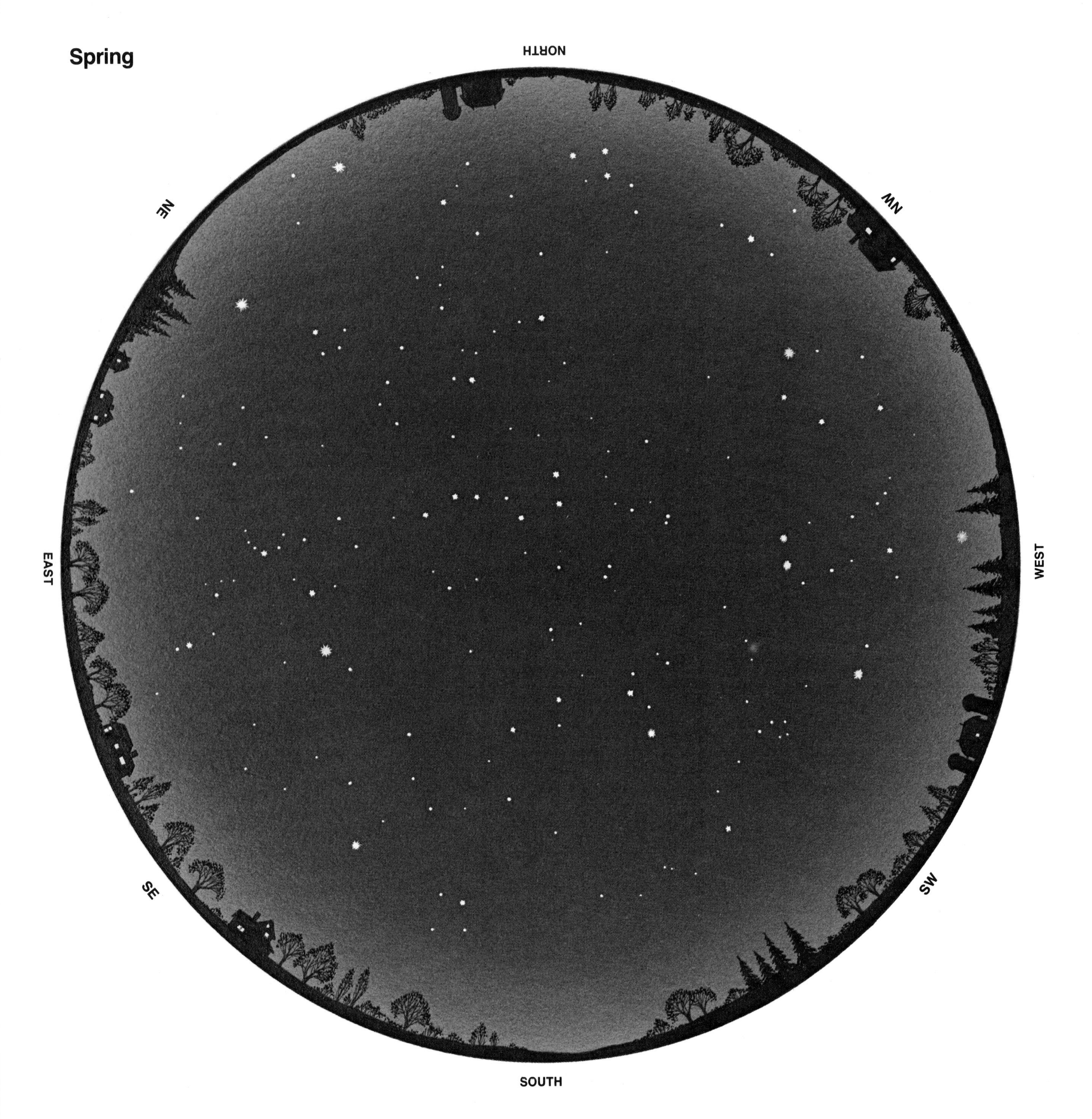

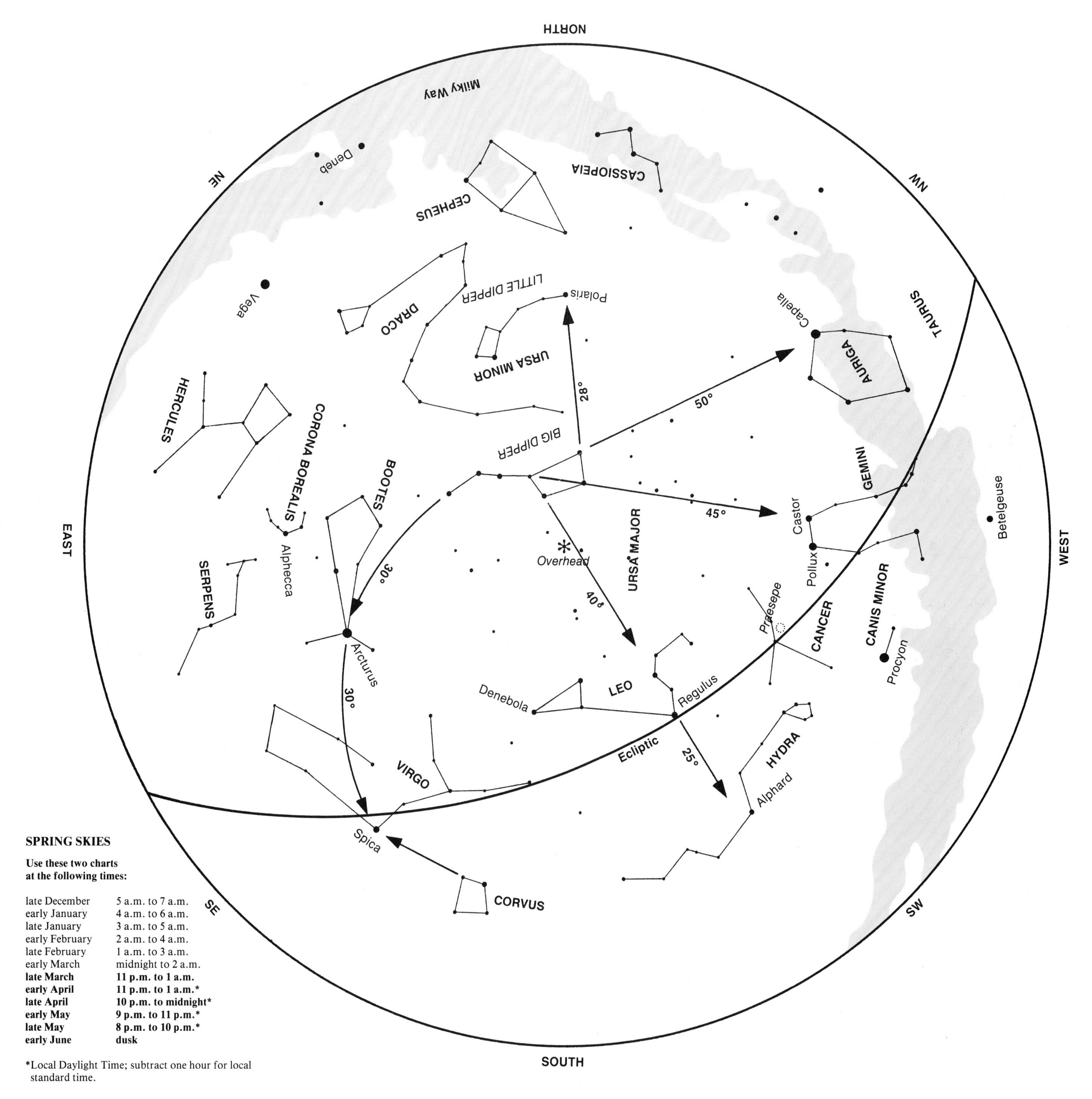

SPRING SKIES

Use these two charts at the following times:

late December	5 a.m. to 7 a.m.
early January	4 a.m. to 6 a.m.
late January	3 a.m. to 5 a.m.
early February	2 a.m. to 4 a.m.
late February	1 a.m. to 3 a.m.
early March	midnight to 2 a.m.
late March	**11 p.m. to 1 a.m.**
early April	**11 p.m. to 1 a.m.***
late April	**10 p.m. to midnight***
early May	**9 p.m. to 11 p.m.***
late May	**8 p.m. to 10 p.m.***
early June	**dusk**

*Local Daylight Time; subtract one hour for local standard time.

Summer

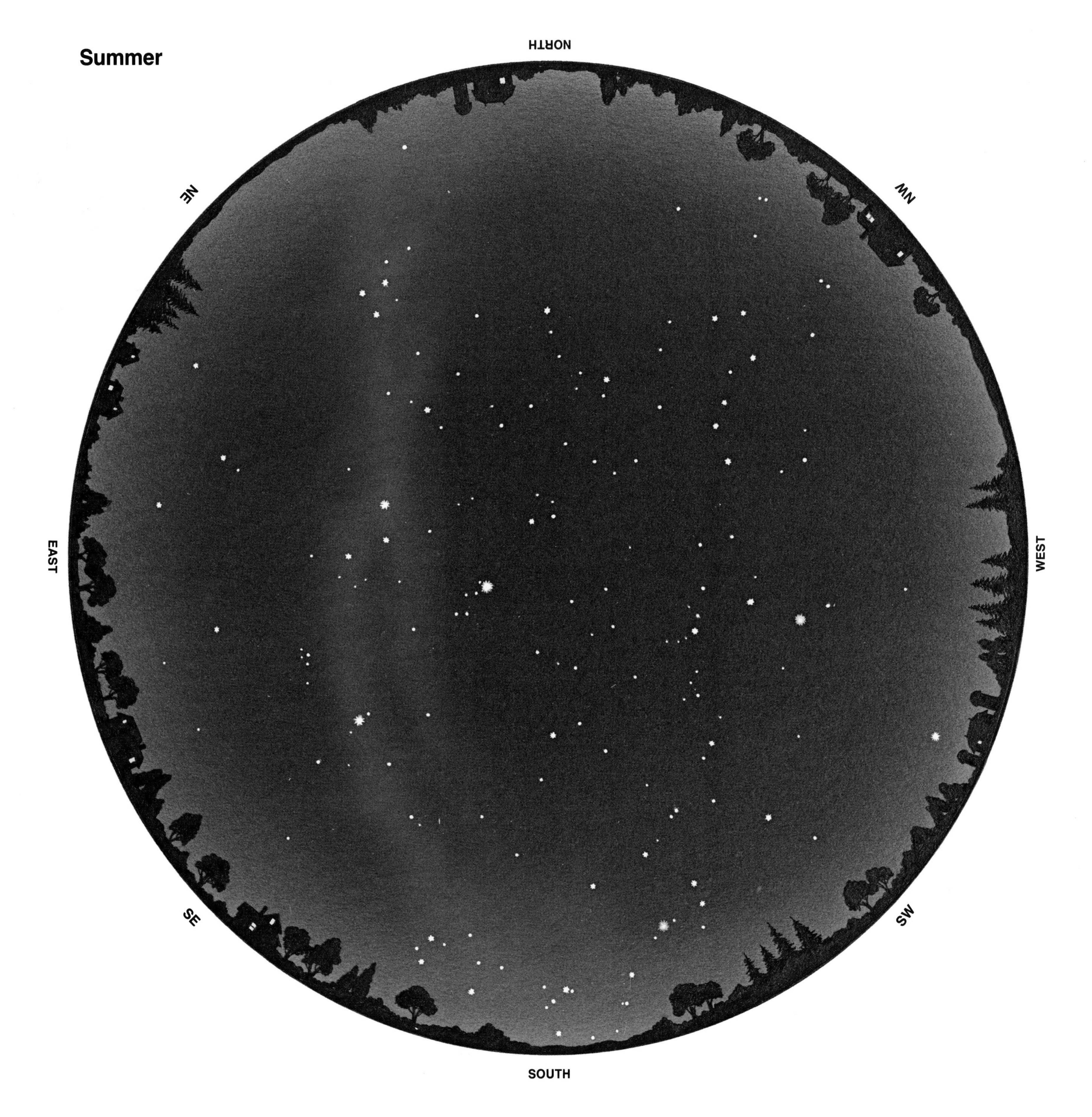

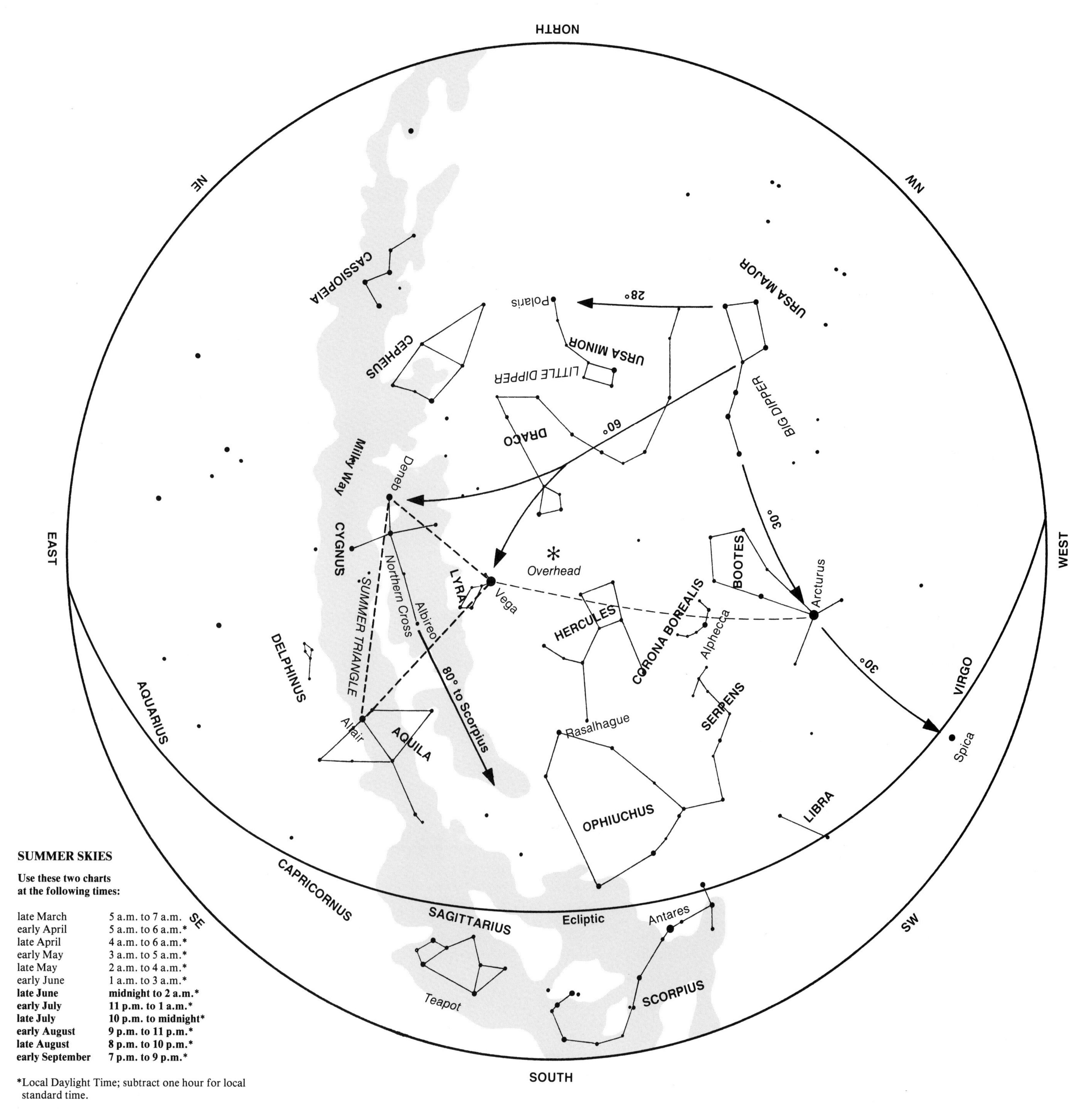

SUMMER SKIES

Use these two charts at the following times:

late March	5 a.m. to 7 a.m.
early April	5 a.m. to 6 a.m.*
late April	4 a.m. to 6 a.m.*
early May	3 a.m. to 5 a.m.*
late May	2 a.m. to 4 a.m.*
early June	1 a.m. to 3 a.m.*
late June	**midnight to 2 a.m.***
early July	**11 p.m. to 1 a.m.***
late July	**10 p.m. to midnight***
early August	**9 p.m. to 11 p.m.***
late August	**8 p.m. to 10 p.m.***
early September	**7 p.m. to 9 p.m.***

*Local Daylight Time; subtract one hour for local standard time.

Turn circular star charts so that the direction you are facing is at chart's lower edge. At night, use a heavily filtered flashlight to illuminate chart so that eyes remain dark-adapted. **Right:** *Dark country skies encroached upon by light pollution from a small city 20 miles away.*

On a very dark night in spring, a beautiful sprinkling of stars (not shown on these charts) can be seen about midway between Regulus and Bootes. Called Coma Berenices, this is a star cluster some 250 light-years distant — very close for a star cluster. (The only other naked-eye cluster that is closer is the Hyades, in the constellation Taurus.) Fewer than a dozen of the cluster stars can be seen with the unaided eye. But if binoculars are turned on this object, a dozen or two more are splashed against a memorable array of background stars.

Praesepe, a star cluster located midway between Regulus and Pollux, should be among the first specific binocular targets in spring skies. It is identified on the colour chart as a pale smudge, just as it appears to the unaided eye. Sometimes called the Beehive, or the Manger, this cluster is a more compact and distant version of the Coma Berenices cluster.

Binoculars will also reveal the star Alcor, a companion to Mizar, the star at the bend of the Big Dipper's handle. The two travel together in space, about a quarter of a light-year apart. Alcor can be spotted without optical aid, but binoculars make it easy. Many more binocular objects are plotted and described on the charts in Chapter 6.

THE SUMMER SKY

If astronomy is good for the soul, then summer is the time for meditation. Summer nights away from the city offer the prime opportunity for discovering the stars. Summer stars in rural skies are so bright, they somehow seem closer, more accessible. With the tranquillity of the night and the majesty of the stellar array, stargazing becomes almost hypnotic, like watching flames in a fireplace.

Exploring summer skies is intimately linked to the Summer Triangle, a large, distinctive figure that is strictly a modern concoction. The triangle comprises Vega, Deneb and Altair, the brightest stars of three separate constellations. They are so much brighter than their neighbours that the triangle is the dominant summer and early autumn configuration.

As usual, the place to start is the Big Dipper, now in the northwest. The two bowl stars nearest the handle point out of the open end of the bowl, about 60 degrees, to an area almost midway between Vega and Deneb. The two are quite easy to distinguish from one another: Vega, at zero magnitude, is noticeably brighter than first-magnitude Deneb. The triangle is completed by Altair, which is also first magnitude but somewhat brighter than Deneb. The Summer Triangle covers a patch of sky slightly larger than the area masked by a hand held at arm's length, fingers spread.

The most prominent of the constellations associated with the Summer Triangle is Cygnus, its main stars forming a cross with Deneb at its top. Popularly known as the Northern Cross, Cygnus is a swan in mythology, its tail at Deneb and its wings stretching beyond the cross's arms. The swan's neck extends to the foot of the cross, the third-magnitude star Albireo. Vega's constellation is the small but distinctive Lyra, the lyre. Altair is the brightest star in Aquila, the eagle, a collection of third- and fourth-magnitude stars with a vague birdlike outline.

With the Summer Triangle's three stars comfortably verified, extend a sight line from Vega to equally bright Arcturus in the west, the major bright star of spring skies. (Arcturus can be confirmed by using the locater arrow that curves outward from the Big Dipper's handle.) This Vega-Arcturus line passes directly through the constellations Hercules and Corona Borealis, the northern crown. Corona Borealis is a small but conspicuous arc of third- and fourth-magnitude stars set off by second-magnitude Alphecca.

The stars of Hercules are more dispersed and less easily distinguished. The Vega-Arcturus line passes through a quadrilateral composed of third- and fourth-magnitude stars, probably the constellation's most distinctive feature. Hercules is the sky's fifth largest constellation in terms of its officially allotted area in square degrees — only Hydra, Virgo, Ursa Major and Cetus are larger — yet its sprawling territory does not include a single star brighter than third magnitude.

Hercules is not alone as a large, faint constellation. Sometimes a constellation has so few bright stars that the region in which it is located seems barren. Such is the case with Ophiuchus, a group whose brightest star, Rasalhague, is about the same magnitude as the central star in the Northern Cross. A direction line running 50 degrees from Deneb through the Summer Triangle, just skimming the southern end of Lyra, leads to Rasalhague. Misidentification is unlikely, since Rasalhague is the brightest star between the Summer Triangle and Scorpius. Ophiuchus has all its reasonably bright stars at its periphery, leaving a vast blank zone.

The brightest star in the southern portion of the summer skies is Antares, in the fishhook-shaped constellation Scorpius, which barely scrapes above the southern horizon on summer evenings. A sight line to Antares starts at Deneb, runs through the Northern Cross's long arm and extends about 80 degrees farther. Antares is a distinctly orange star, brighter than Altair but fainter than Vega. (The locater arrow from the Northern Cross aims almost directly at Antares. Unavoidable distortion in all-sky maps produces the offset along with the slight curve in the Vega-Arcturus line, which is actually straight in the real sky.)

In Arabic, Antares means rival of Mars, and the name fits. When the reddish planet rides the ecliptic in this region, the two look almost identical. Antares' red-orange colour originates with the star. Only half the average temperature of our sun, Antares is one of the rare red supergiant stars, with a diameter about 500 times as great as the sun's. If Antares replaced the sun, it would enclose the Earth's orbit. If this star were at Vega's distance of 27 light-years, instead of 500, it would be magnitude −6, by far the brightest object in the sky next to the moon.

Also riding low in southern summer skies is the teapot-shaped constellation Sagittarius. The teapot's spout is to the right, its handle to the left. Identify this constellation by running a sight line from Deneb through the Summer Triangle, just to the right of Altair, and extending it to near the south horizon.

The Summer Triangle stars demonstrate the amazing variety of suns that populate the Milky Way Galaxy. Altair, second brightest of the three, is the nearest, some 17 light-years away. Altair is basically like the sun but about 10 times brighter. To the unaided eye and in binoculars, Altair's colour is white, similar to the colour of light radiated by our own sun.

Vega, the brightest member of the Summer Triangle, is actually more distant than Altair. Astronomers estimate that Vega is 58 times brighter than the sun. This is partly because Vega's surface temperature is twice as hot as the sun's. Higher temperature means that more energy is released per unit area and that the star shines bluish white.

To the unaided eye, the difference between Altair and Vega should be noticeable — Altair's white contrasting with Vega's bluish white. For a more obvious colour contrast, compare Arcturus (yellow) and Antares (orange) to Vega. The hottest stars are blue, the coolest reddish orange. Deneb, the dimmest star in the Triangle, is the same bluish white colour as Vega but, in reality, is by far the brightest of the three.

Deneb is so far from the sun — 1,600 light-years — that astronomers are uncertain how bright it is, but their best estimates place it at 60,000 times the power output of our sun. Such an amazing superstar so dwarfs the sun that if Deneb were to be in its place, Earth could be as far away as Pluto and still receive five times more heat and light than it now does.

The guideposts just described provide a foundation to build on when one is seeking less prominent stars and constellations of summer. This is the key to the star-and-constellation identification technique. Always start with the brightest, most obvious stars, linking them by locater lines, arcs and triangles to surrounding objects, and then fill in the details.

One of the joys of amateur astronomy is scanning the summer Milky Way through binoculars. Arching across the summer sky from northeast to southwest, the pale misty band is best seen from rural locations when the moon is absent. What is seen with the unaided eye as a cloudlike ribbon is transformed into a glittering river of thousands and thousands of stars.

The Milky Way appears misty to the unaided eye because of the eye's inability to resolve it into its individual stars. Actually, we are looking through the densest parts of the Milky Way Galaxy to the other spiral arms beyond our own. The centre of the galaxy is very close to the tip of the spout on the Sagittarius teapot, but 300 times farther away. The Milky Way is particularly rich in this vicinity, although it would be several thousand times brighter if dense clouds of gas and dust did not obstruct our view to the nucleus.

After some general binocular exploration, noting the rifts and clouds of stars in the Milky Way and the relative lack of

Autumn

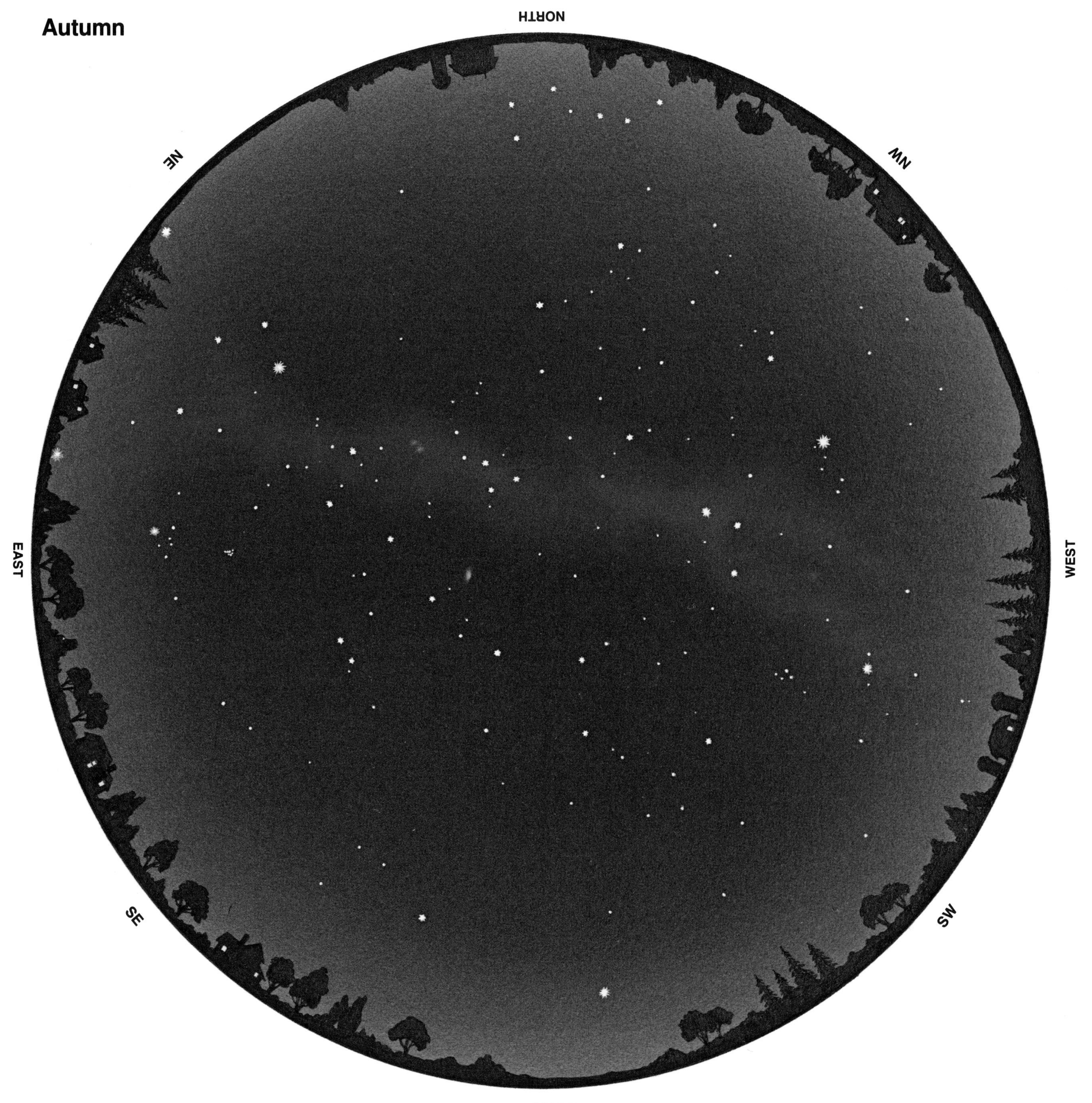

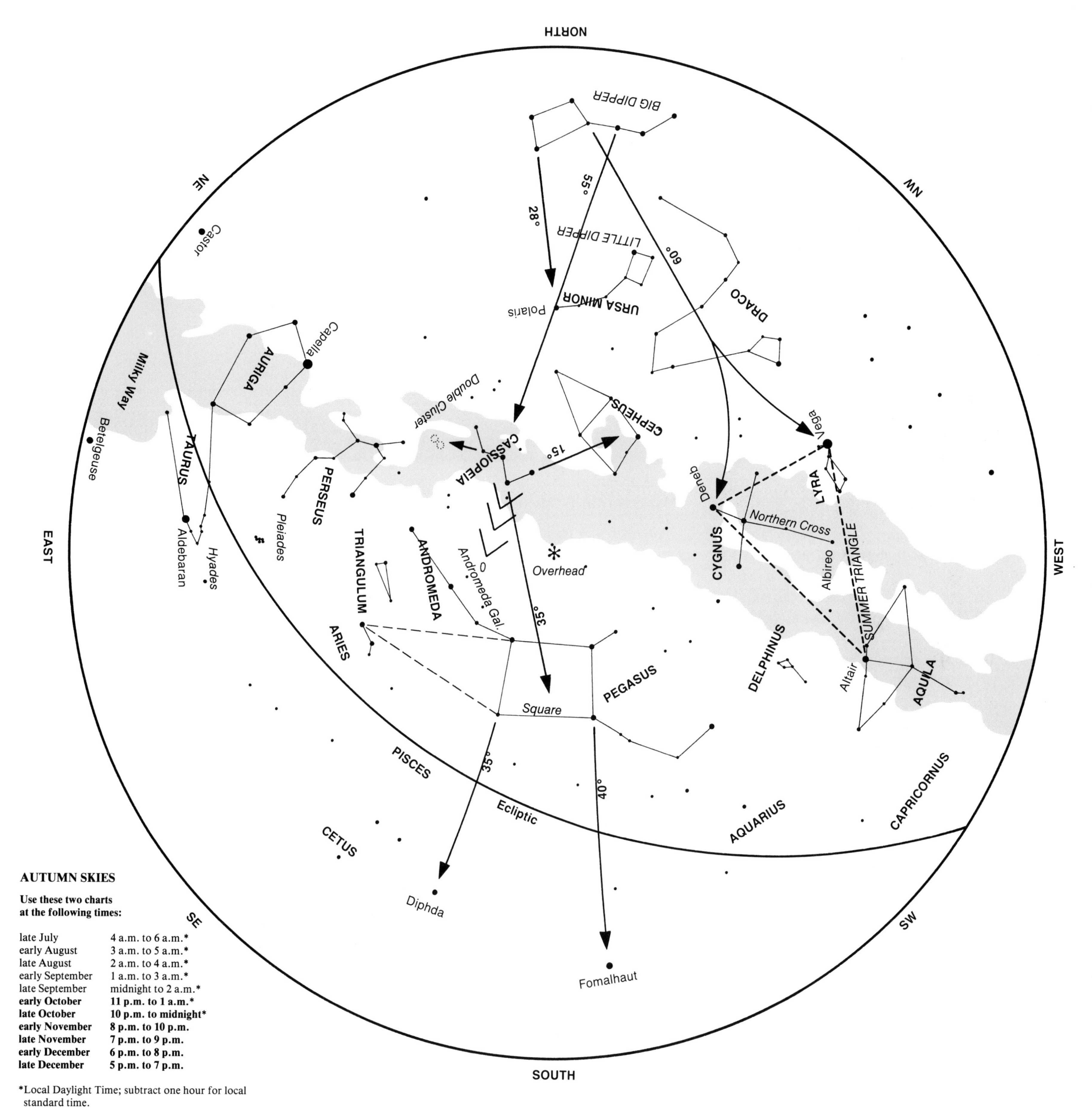

AUTUMN SKIES

Use these two charts at the following times:

late July	4 a.m. to 6 a.m.*
early August	3 a.m. to 5 a.m.*
late August	2 a.m. to 4 a.m.*
early September	1 a.m. to 3 a.m.*
late September	midnight to 2 a.m.*
early October	**11 p.m. to 1 a.m.***
late October	**10 p.m. to midnight***
early November	**8 p.m. to 10 p.m.**
late November	**7 p.m. to 9 p.m.**
early December	**6 p.m. to 8 p.m.**
late December	**5 p.m. to 7 p.m.**

*Local Daylight Time; subtract one hour for local standard time.

stars elsewhere in the sky, there are some specific binocular sights worth pursuing. In the constellation Lyra, for example, binoculars will reveal the star closest to Vega, in the direction of Deneb, as a beautiful twin star. (In fact, people with sharp eyesight will see these two stars without any optical aid.) This is Epsilon (ϵ) Lyrae, a fascinating star system that is detailed on Chart 10 in Chapter 6.

Another region that should be on any binocular sky tour is the rich zone between the end of the fishhook of Scorpius and the spout of the Sagittarius teapot. Especially beautiful are two clusters of dozens of stars, like swarms of fireflies in the night. (Unfortunately, this star-studded region is too close to the southern horizon for a decent view from north of about 48 degrees latitude.)

Binoculars also enhance the view when one is observing from the city. Just as in the country, where they can show stars fainter than those visible to the unaided eye, in the city, binoculars reveal stars that may be suppressed by smog and artificial lights. The stars of Lyra, for instance, show up clearly in binoculars from urban locations when only Vega is visible to the unaided eye.

THE AUTUMN SKY

Autumn's long evenings and generally comfortable weather for observing combine to produce perfect conditions for backyard skygazing. In June and early July, it is often not dark enough for convenient skygazing until 10:00 in the evening, whereas in October, two extra hours of evening darkness permit leisurely investigation of the night sky.

The autumn sky contains fewer bright stars and generally less distinctive stellar configurations than can be seen during the other three seasons. However, compensation is afforded by more than a dozen second-magnitude stars, which form easily recognized star groups harbouring some of the sky's greatest wonders.

The Big Dipper scrapes low toward the northern horizon during autumn evenings, so a dark sky and an unobstructed northern horizon are prerequisites for use of the Big Dipper's locater-arrow system. The key autumn locater arrow is the one that emerges from the third star in the Dipper's handle and passes through Polaris to Cassiopeia, near overhead, a total of about 55 degrees. If the sight line from the Big Dipper is obscured by trees or lights in the north, scan the overhead region for the distinctive W-shape of Cassiopeia. The constellation is about 15 degrees wide, with each arm of the W three to four degrees long. Cassiopeia, the mythological queen, governs autumn-sky identification. No fewer than four locater arrows emanate from this small configuration, the most important extending south about 35 degrees to the centre of the "square" of Pegasus.

The square is fairly large — its sides range from 14 to 17 degrees in length and are marked by four second-magnitude stars. When the observer faces south, the right side of the square forms a sight line south to Fomalhaut, a first-magnitude star near the horizon. The left side of the square similarly aims down to a second-magnitude star, Diphda, in the gigantic but dim constellation Cetus. The nearly blank southeastern quadrant of the sky is what I call the Cetus void. Of all the areas in the sky that do not have any first- or second-magnitude stars, this is one of the largest. The top margin of the Cetus void contains the tiny zodiac constellation Aries, whose brightest star, Hamal, is at the end of an isosceles triangle connected to the eastern side of the square.

The star in the square closest to Cassiopeia does not officially belong to Pegasus. It is part of the constellation Andromeda, whose stars angle up to the northeast. Andromeda's astronomical claim to fame is that it contains the only galaxy similar to the Milky Way Galaxy which is visible to the unaided eye. (Two small satellite galaxies of the Milky Way, the Large and Small Magellanic Clouds, can be seen without optical aid, but only from south of the equator.) The triangle of stars forming the half of the W closest to Pegasus acts as an arrowhead pointing southward 15 degrees to the Andromeda Galaxy.

The Andromeda Galaxy is a faint fourth-magnitude smudge, the most distant object that can be viewed with the unaided eye, but moonless, dark skies are needed to see it. It is almost precisely overhead on November evenings, appearing like an erasure mark on a blackboard. At times, it seems not to be there at all, tantalizingly at the threshold of vision. This fragile, hazy patch is an enormous swarm of suns so remote that the combined energy of its 500 billion stars barely produces a detectable image in the eye.

Queen Cassiopeia's king is Cepheus, whose third- and fourth-magnitude stars form the shape of a kindergartner's drawing of a house. Between Cassiopeia and the adjacent constellation, Perseus, is a neat twin star cluster called the Double Cluster. Slightly easier to see with the unaided eye than the Andromeda Galaxy, it is a similar hazy patch. The dual nature of the Double Cluster is evident in binoculars, but its individual stars are difficult to detect without a telescope.

The double cluster is 7,000 light-years away, in the Milky Way spiral arm beyond our own. Cassiopeia and Perseus are in almost the opposite direction from the centre of the galaxy; but the Milky Way is still rich and impressive in this region.

STARS OF WINTER

It is often said that the stars shine more brightly on crisp, clear winter nights than at any other time. Although the stars may seem brighter, actual measurements prove that there is no difference in clarity between the best skies in winter and those at other times of the year.

Triangulum Galaxy (M33), **far left**, *is the Milky Way's second closest neighbour spiral galaxy. At a distance of 2.5 million light-years, it is just below naked-eye visibility. Two fainter objects, the Horsehead Nebula,* **above**, *and NGC 1977,* **left**, *are clouds of interstellar dust and gas illuminated by nearby stars. Both are located within the Milky Way Galaxy.*

However, there is another factor to consider. A cold winter night means a significant drop in temperature from daytime. This cooling causes turbulence in the Earth's atmosphere, which manifests itself by making the stars appear to twinkle as the rippling air moves in front of their fragile beams of light. This twinkling tends to draw more attention to the stars in winter skies.

But there is one major difference in the winter sky that makes it intrinsically brighter. The *number* of bright stars visible is greater than at other seasons. There are more stars shining, so in that sense, winter has brighter nights.

Many of these cold-weather celestial luminaries are contained in the constellation Orion, the brightest of all the classical star groups and what is generally regarded as the most distinctive stellar configuration in the heavens after the Big Dipper. Unlike most constellations, which bear no resemblance to their namesakes, Orion the hunter actually looks like a hunter. The nimrod's unmistakable three-star belt is unique. Nowhere else in the sky are three stars of this brightness so close together. Surrounding the belt are four stars marking Orion's shoulders and legs.

Rigel, the brightest star in Orion, is one of the most luminous stars known. Shining about 55,000 times more powerfully than the sun, this hot blue-white stellar beacon is 900 light-years distant. More than a million stars are closer to us than Rigel, but not one of them can match its mighty energy output.

The second brightest of Orion's suns, Betelgeuse, is equally impressive, since it is one of the largest stars known. With an estimated diameter about 800 times greater than the sun's, Betelgeuse would easily enclose the orbits of Mercury, Venus, Earth and Mars if it were to replace our sun. It is a member of an exclusive class of rare stars known as red supergiants — obese stellar spheres of low temperature. To the unaided eye, Betelgeuse is distinctly ruddy.

Betelgeuse and the rest of Orion can be found any clear winter evening almost due south, about halfway from the horizon to overhead. The belt is about three degrees wide, and the apparent distance from Rigel to Betelgeuse is just under 20 degrees. If the line formed by Orion's belt is extended 20 degrees down to the left, it leads to Sirius, the brightest star in the night sky, at magnitude −1. When the air is turbulent, the star appears to twinkle violently, sometimes changing colour to blue, yellow or white, offering an almost unending display, like a glittering diamond. When the air is steady, Sirius should have a bluish white tinge.

Orion's belt points in the opposite direction an equal distance (20 degrees) to Aldebaran, a first-magnitude star with a definite yellowish orange tinge. If this

Winter

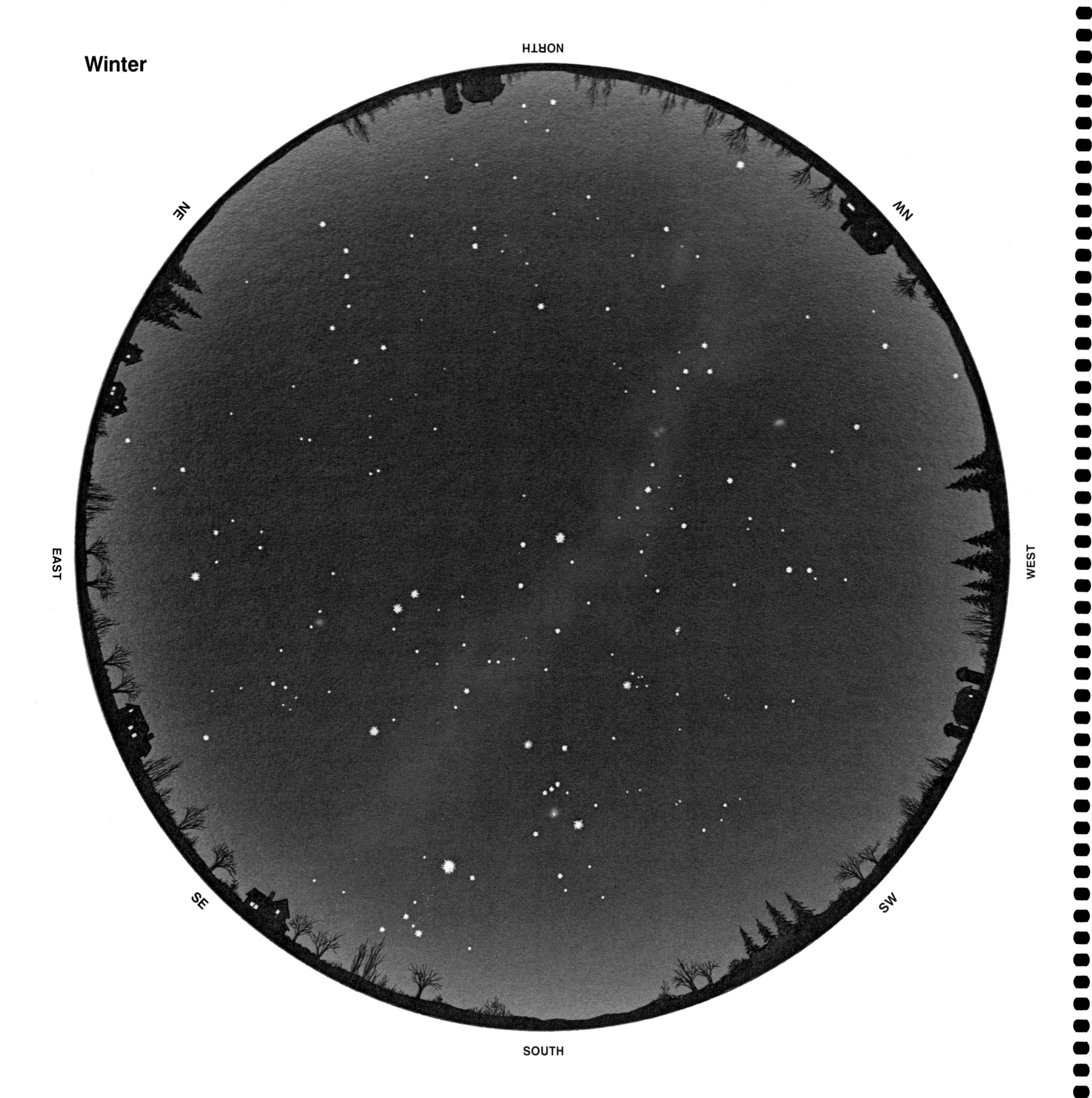

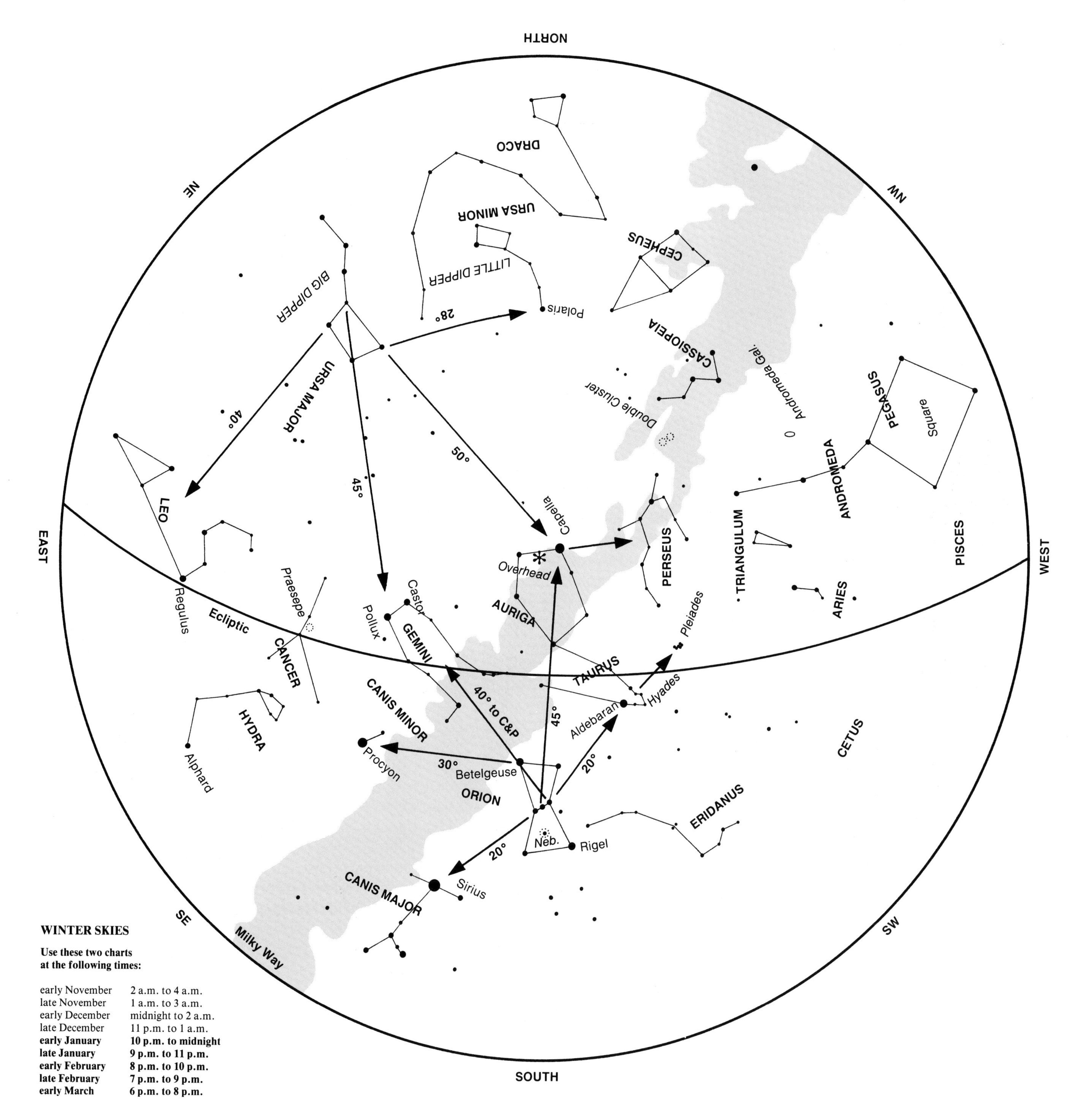

WINTER SKIES

Use these two charts at the following times:

early November	2 a.m. to 4 a.m.
late November	1 a.m. to 3 a.m.
early December	midnight to 2 a.m.
late December	11 p.m. to 1 a.m.
early January	**10 p.m. to midnight**
late January	**9 p.m. to 11 p.m.**
early February	**8 p.m. to 10 p.m.**
late February	**7 p.m. to 9 p.m.**
early March	**6 p.m. to 8 p.m.**

pointer from the belt to Aldebaran is continued another 15 degrees, it leads to a beautiful cluster of stars known as the Pleiades, the seven sisters.

The Pleiades is the brightest and the most distinctive star cluster in the sky. The two-degree-wide cluster is a wonderful sight in binoculars. Because of its shape, it is often incorrectly called the Little Dipper. The real Little Dipper (Ursa Minor) is near the Big Dipper and, in any case, is much less prominent than the Pleiades.

A line running from the middle of Orion's belt straight north through the top of the constellation, between Betelgeuse and Bellatrix, and extending on for some 45 degrees to a point almost exactly overhead comes to Capella, a star second only to Sirius in brightness in the winter sky. From Capella to Sirius, a huge, gently curving arc can be traced that touches three other bright stars: Castor, Pollux and Procyon. These stars also have locater sight lines from Orion.

The diversity of the winter group of bright stars, unmatched in other seasons, provides a good example of the tremendous variation in the celestial zoo of stellar types. Sirius, the brightest, is about nine light-years away, one of the closest of all stars, and is the nearest star visible from Canada and the United States. It is bigger and brighter than the sun, with about twice the sun's diameter and 23 times its brilliance. Sirius's exceptional brightness is due to its substantially greater energy output and hotter temperature compared with our sun.

Capella, Procyon, Rigel and Betelgeuse are all zero-magnitude stars and appear to be almost the same brightness to the unaided eye. And yet this similarity is sheer coincidence. Procyon is only 11 light-years away, Capella is 45 and Betelgeuse is 520, while Rigel is a colossal 900 light-years distant.

Next in order of brightness in the winter group comes Aldebaran, a first-magnitude star classified as a red giant – a smaller version of Betelgeuse. Aldebaran is 360 times the sun's brightness and 45 times its diameter and has a distinct orange cast similar to that of Betelgeuse.

Still working through the bright stars in the winter group, we come to Pollux, a first-magnitude star 35 light-years away and 33 times the sun's luminosity. Its companion, Castor, is officially classified as a second-magnitude star but is almost as bright as Pollux. Castor is 45 light-years away. Because they are so close in apparent brightness, it is often difficult to know which is which. I do it by remembering that Castor is closest to Capella and Pollux is closest to Procyon. The alliteration provides the mnemonic device.

Castor and Pollux are the brightest members of Gemini, the twins. Two chains of stars extending toward Orion from each of these stars form the mythological twin boys, a designation of little help when searching for the star groups. Gemini is famous because it is part of the zodiac, the 12 constellations that happen to be located on the ecliptic. Centuries ago, astrologers (not to be confused with astronomers) first attached importance to these constellations because the moon, planets and sun appear to move through them.

A second zodiac constellation in the winter group is Taurus, the bull, the bright star Aldebaran marking one eye of the beast. Aldebaran is at the end of one arm of a small V of stars plainly visible in moderately dark skies. This is the Hyades, a star cluster similar to the Pleiades but more spread out and therefore less striking. Binoculars reveal dozens of stars just below the limit of naked-eye visibility.

A unique binocular object is nestled underneath Orion's belt and is indicated on the chart by "Neb." This is the great Orion Nebula, the brightest and nearest of thousands of such clouds scattered throughout the galaxy's spiral arms. To the unaided eye, it is definitely a small, fuzzy patch, rather than a sharp starlike point. Binoculars confirm this impression, showing the teacup shape of this extended cloud – a puff of cosmic gas seemingly frozen in timeless space.

The Milky Way, that misty band of light which is usually associated with summer skies, extends almost entirely across the winter sky from northwest to southeast. The Milky Way's winter sector is not quite as bright as its summer counterpart.

Some of the best parts of the winter Milky Way are between Orion and Gemini and up through Auriga. Binoculars expose the true stellar nature of this celestial ribbon of light. Thousands of stars pass through the field of view in places where only a pale haze is seen with the eye.

Swaths of the Milky Way visible in summer, autumn and winter skies act as a reminder that we observe the universe from within a disc of stars. The stellar scene probably was not much different a million, or a billion, years ago because the sun's path through space keeps it in the vicinity of the spiral arms. As timeless as the vista may be, only in the last few decades has anything close to a comprehensive picture of the cosmos come into focus. The creatures who inhabit one planet of one star of one galaxy among billions now know the extent, if not the ultimate meaning, of the visible universe. Late-20th-century backyard astronomers peer out into the void with knowing eyes.

Left: *The Hyades and Pleiades star clusters are seen in detail in this photograph, along with the brilliant planet Jupiter below the Pleiades. The bright yellow star at lower left is Aldebaran. All stars shown are visible in binoculars.*

THE TRAGEDY OF LIGHT POLLUTION

Children growing up in the final decades of the 20th century represent the first generation in the history of civilization to live in a world where the stars are almost certain to be the last thing noticed at night instead of the first. The change has been swift. Anyone in their fifties or sixties can clearly remember when the splendour of a dark night sky dusted by the pale glow of the Milky Way was as close as the back door.

Today, ubiquitous outdoor lighting is a fact of life, an accepted and ignored part of urban existence. Yet all it takes is one attempt to identify a few stars and your awareness of night lighting abruptly sharpens. From within a city (large or small), you soon realize the sky is grey, not black. Outdoor lighting illuminates the air as well as the ground.

To demonstrate the extreme difference between city and country starscapes, I took the two pictures on this page just a few days apart using the same camera, lens, film and exposure (25 seconds). The identical sector of the sky is shown. The sky was moonless and exceptionally clear in both cases; the only difference was location. One photograph was taken in rural Ontario, well away from metropolitan areas. The other is a view from my mother-in-law's condominium, which faces Toronto (population three million) from near the edge of the city. (Observant readers will have already noticed the motion of the planet Mars between photographs.)

Apart from general urban sky glow, every stargazer comes to hate the one or two local lights that seem to shine directly in the eyes. The offending luminaire is usually a streetlight, but porch lamps and "security" lighting are often the source as well.

Since many people spend as little time as possible standing outside at night, they never notice that most outdoor lighting produces this glare. It arises from either poorly designed or poorly installed fixtures that pump light in all directions instead of limiting it to the intended target. Rarely is there a need to direct *any* nighttime illumination horizontally or higher. It is pure wasted energy — light pollution.

Light pollution in North America amounts to approximately one billion dollars a year. That's money spent to generate electricity for light that never touches the ground but instead uselessly illuminates the night sky and robs us of a clearer view of nature.

What can you do? Invite your friends or neighbours to look through your telescope. Once they turn their eyes skyward, the problem of glaring lights often becomes self-evident. Probably the most meaningful impact you can have is to set an example by switching your own outdoor lights to fixtures with an infrared motion detector. They turn on only when something moves in the vicinity. These systems save energy and are much more likely to deter someone nasty than a steady glow. For your own use, they come on when you step outside or arrive home and can be overridden by the normal light switch when you are outside observing.

City and country views — identical exposures with the same camera — dramatically demonstrate the tremendous amount of light thrown into the night by streetlights and other sources of urban illumination.

THE ECLIPTIC AND THE ZODIAC

Although we have dispensed with grids and celestial coordinates on our charts, one vital line remains: the ecliptic, the boulevard of the celestial wanderers — the sun, the moon and the planets. In an oversimplified sense, our solar system is like the surface of a wide, circular racetrack, and the planets are the race cars. Sometimes Venus overtakes Earth, sometimes Earth passes Mars, but all the action happens in the same horizontal plane. Thus the planets, sun and moon are always seen in a restricted band in the sky corresponding to this plane.

The reason for the solar system's flatness can undoubtedly be traced back to its origin, about five billion years ago, in a vast pizza-shaped cloud of cosmic dust and gas. Regardless of its genesis, this system is so flat that the moon and planets are seldom more than a few degrees away from the ecliptic.

The stars that form the backdrop for the ecliptic are known as the zodiac, a band of 12 constellations made familiar by the ubiquitous newspaper horoscopes. The word zodiac derives from the Greek *zodiakos kyklos*, meaning circle of animals. However, there are only 7½ animal signs among the 12 constellations: Aries the ram, Taurus the bull, Cancer the crab, Leo the lion, Scorpius the scorpion, Capricornus the goat and Pisces the fish. The half-and-half constellation is Sagittarius the archer — half man, half horse. Then there are human figures: Aquarius the water carrier, Gemini the twins and Virgo the maiden. Of course, these are animals too, but Libra the balance does not fit into even the loosest animal definition. Libra seems to be a later revision intended to distinguish this region of the sky from its former designation, the claws of the scorpion.

Some of the zodiac constellations are faint, despite their important location on the pathway of the planets. To coordinate with the planet tables on the inside back cover, every zodiac constellation is named on the all-sky charts, although some of them are purposely not shown to avoid giving the impression that they are as distinctive as the other star groups. However, all zodiac constellations are shown in the detailed charts in Chapter 6.

The zodiac constellations are exceedingly ancient, having originated as early as 3300 B.C. Drawings on Mesopotamian artifacts from that time depict Leo and Taurus in combat. Clearly, the Mesopotamian artists were portraying the constellations, because the drawings are adorned with star symbols. The late Michael Ovenden of the University of British Columbia, an expert on the origin of the constellations, deduced that most of the zodiac was designed about 2600 B.C. Far from being invented by shepherds and nomads as an amusing pastime, the constellations of the zodiac were carefully selected to define and describe, in a useful way, the positions of the sun, moon and planets.

Having one practical celestial mapping system would have been of vital importance to ancient sailors. The most astute sailors of 4,500 years ago were the Minoans of Crete, who may have convinced other cultures of the value of a single standardized code of sky geography. The Greeks, and eventually the Romans, refined the system into the form still used today, complete with its mythology and names.

The original division of the zodiac into 12 constellations could have emerged when early skywatchers noticed that Jupiter requires 12 years to complete its trip around the ecliptic, spending one year in each zodiac constellation. Jupiter is the brightest planet seen throughout the night and must therefore have been an object of great interest. (Venus is brighter but is visible for only a few hours before sunrise or after sundown.) Far back in the mists of antiquity, the number 12 became a powerful symbol — 12 apostles, 12 biblical patriarchs, 12 jurors and, of course, 12 months in a year. This final division is most likely a product of the lunar orbit: the moon sweeps around Earth 12 times a year, with 12 days left over.

The specific groups of stars associated with each zodiac constellation vary greatly in size. Virgo, the largest, is three times the size of Aries, the smallest. Astronomers are unconcerned about the discrepancies that arise from the stellar configurations developed so long ago. But ancient astrologers created the *signs* of the zodiac, each 30 degrees wide, to overcome these inequalities. Modern-day astrologers refer to the zodiac signs in horoscopes and elsewhere when they mention Taurus, Gemini and the others. They are *not* referring to the constellations. Two thousand years ago, the signs and the constellations were approximately aligned, but a slow oscillation of the Earth's axial orientation has shifted the constellations, relative to the seasons, west, so that today, the signs do not coincide with the constellations. When an astrologer says that Mars is in Gemini, for example, it is really among the stars of Taurus. Astrology is based on a system that no longer reflects the reality of the sky.

Straddling the Milky Way, the zodiac constellations Scorpius and Sagittarius bear a closer resemblance to a fishhook and a teapot than to their mythological namesakes. Scorpius is at right; Sagittarius, lower left. *(See page 95.)*

5

Stargazing Equipment

O, telescope, instrument of much knowledge, more precious than any sceptre, is not he who holds thee in his hand made king and lord of the works of God?

— Johann Kepler

Armed with nothing more than a pair of 25 x 105 binoculars, English amateur astronomer George Alcock has found four novas from his backyard in Cambridgeshire. (A nova is the sudden brightening of a formerly obscure star.) Nova discoveries by amateurs are quite rare — only two or three per decade — and Alcock's feat remains a world record.

Alcock, now retired from the British postal service, has also discovered five comets, two of them with smaller 15 x 80 binoculars. He began his sky scanning in 1955, using 12 x 40 bird-watching binoculars, and slowly graduated to the larger models because they show fainter stars.

In the late 1950s, he decided that the most efficient way to search for comets and novas would be to memorize the positions of all the stars visible in his binoculars, thus avoiding time-consuming referrals to a star atlas. Every clear night for the past 30 years, Alcock has spent hours reclining in a deck chair in his backyard, patrolling the sky with binoculars.

Etched in his memory are the positions of at least 30,000 stars, down to eighth magnitude, plus 500 nebulas and galaxies that can masquerade as comets. As he examines the sky, he checks off what he sees against his mental star map, like a computer searching its memory banks. Any interlopers among the stars are recognized instantly.

Alcock can claim to know the sky better than anyone in the world, perhaps in history. Yet his astronomical equipment consists solely of binoculars.

His most impressive feat came the evening of May 3, 1983. While sitting in his kitchen during a respite from the dampness outdoors, Alcock began scanning the sky through the closed kitchen window with his 15 x 80 binoculars. As he worked his way through a familiar star field in the constellation Draco, he spotted a fuzzy patch that he immediately suspected was a comet. It was.

His fifth comet discovery, officially known as Comet IRAS-Iraki-Alcock, passed closer to Earth than any comet since 1770 and was observed by thousands of skywatchers. Astronomers considered the chance to watch a comet race past Earth a once-in-a-lifetime opportunity. And it all started with binoculars.

Next to a camera, binoculars are probably the most common optical equipment in the home. Yet I am always surprised to find how seldom, if ever, people turn their binoculars skyward. The sky's tapestry of stars becomes a pageant of jewelled velvet in even the most modest binoculars. The colour of stars also tends to be more intense in binoculars than when viewed with the eyes alone.

Binoculars are actually two small telescopes that are more compact than conventional telescopes. The light path is compressed by the use of prisms within the structure of the instrument. Binoculars do not transmit as much light as similar-sized telescopes designed for astronomical purposes, but they make up for it in compactness and ease of use. The fact that binoculars permit viewing with both eyes is another bonus. In my experience, binocular observing is most comfortably

Binoculars are essential equipment for both beginning and advanced backyard astronomers. **Above:** *A comparison of the main lenses of a 60mm refractor telescope and 11 x 80 and 7 x 50 binoculars.* **Facing page:** *A binocular tripod adapter,* **left,** *steadies the instrument for sharper views. A reclining lawn chair,* **centre,** *is ideal for comfortable scanning with binoculars. However, the ultimate is a custom-made swivel chair,* **right.**

done when reclining in a lawn chair. The less jiggling, the more pleasing the view.

The largest-selling binoculars are the 7 x 35 size, which means that they magnify seven times and have main lenses 35mm in diameter (about 1.4 inches). Thus they are approximately equivalent to a telescope with a 35mm aperture and a 7-power (7x) eyepiece. Although an instrument of this size brings in the star fields of the Milky Way and provides pleasing views of some star clusters, such as the Hyades and the Pleiades, it is noticeably outperformed by 7 x 50 or 10 x 50 binoculars. The 50mm (2-inch) lenses on these larger binoculars have double the light-collecting power of 7 x 35s and show stars almost a magnitude fainter. Typically, 50mm binoculars are able to detect ninth-magnitude stars and reveal all the sights described in this book as suitable for binoculars.

Using higher-power binoculars, however, involves a trade-off. The higher the power, the narrower the field of view. A small field of view means that less is seen at one time and that aiming is more difficult, especially at night. Most binoculars have their fields of view indicated in terms of feet at 1,000 yards. This can be translated into degrees of field of view — a more useful comparison for astronomical purposes — since an angular distance of 52 feet at 1,000 yards is one degree.

The average 7 x 50 binoculars have a field of view of 375 feet at 1,000 yards, or about seven degrees — much larger than any telescope — offering unique and comfortable sky views. On the other hand, 20 x 50 binoculars, with their three-degree field of view, produce such a restricted field that they find little favour among astronomy enthusiasts.

I think there is a strong case for the 7 x 50 and 10 x 50 sizes. They provide the optimum balance between weight, light-collecting power, field of view, ease of handling and adaptability to other recreational uses, such as birding. My second choice would be 7 x 42 or 8 x 42, sizes widely favoured by experienced birders and naturalists.

What should you buy? I have tested hundreds of binoculars and have reached a few conclusions about what to look for and what to avoid. First of all, if you already own binoculars in the 35mm or larger size, try them for astronomy. If they are a discount store's $29.95 special, they will at least get you started, but there are not many binoculars available today for under $100 that are much more than training wheels — serviceable for introductory viewing. In general, $75 binoculars that work fine for boating and casual daytime viewing have deficiencies that become noticeable in stargazing. The most stringent test for optics is imaging bright points on a black background.

Top-quality binoculars produce tiny pinpoint star images over all but the extreme edge of the field of view. They have special lens coatings and baffling that reduce light loss and internal reflections. The lenses and prisms are made to exacting standards and are designed to deliver bright, sharp, aberration-suppressed views. And they are expensive. Even so, I have never seen a truly perfect binocular for astronomy (I'm fussy). Some have superb optics but are too heavy to hold, others are jewels in every respect but have astronomical price tags (up to $2,000). The best binoculars I have seen for under $400 are the Adlerblick series by Carton Optics of Japan. In both 7 x 50 and 10 x 50 sizes, they are exceptionally lightweight and produce very sharp images. Nikon binoculars are also first-class for astronomy. During my tests, I found several binoculars with obscure brand names in the $150 range that were very fine performers. Price seems to be a better guide than familiar brand names in this competitive field. Avoid "zoom" binoculars, because they have little application in astronomy but cost substantially more.

Some binoculars, rated by their manufacturers as "wide angle," yield amazing picture-window vistas up to 11 degrees wide. These instruments give stunning

views of the Milky Way and other star-rich areas of the sky, but usually, significant portions of the outer edge of the field are out of focus due to inherent limitations of wide-angle binocular optics. This is more noticeable in astronomy than in day-time land viewing, and I find the defect objectionable. But the choice is mainly one of personal preference.

When evaluating a binocular for astronomy, a key consideration is the tripod adapter hole – a threaded hole at the front of the central bar. An inexpensive L-shaped bracket fits the hole and allows your binocular to attach to a camera tripod. Significantly more detail – especially subtle astronomical detail – is visible through binoculars steadied in this way. Any binocular purchased for stargazing must have the adapter hole. Many do not, so ask about this feature before you buy. If the dealer does not know what you are referring to, go elsewhere.

Tripods are essential for binoculars over 50mm aperture. They are simply too heavy to hold for more than a few seconds. My 11 x 80s weigh in at just under five pounds, compared with 1½ pounds for my 10 x 50s. The brutes also come in 10 x 70, 16 x 70, 15 x 80, 20 x 80 and several other configurations. When fixed to sturdy camera tripods, they are fine for astronomy and long-distance land viewing. Large binoculars are the transition instruments between hand-held binoculars and telescopes. Large binoculars are optional equipment in my assessment, whereas a good-quality standard-sized binocular is an essential tool for the astronomy enthusiast.

TELESCOPES

Good binoculars and a dozen of the best astronomy books are the bare essentials for the backyard astronomer. Yet such minimal aids can provide years of stimulating exploration – by mind and eye – into the depths of space. But sooner or later, almost everyone who is captivated by the mystery of the starry night craves a telescope. The problem is, the craving usually starts sooner, rather than later. More than a million telescopes have been sold in Canada and the United States over the last 25 years. Most of them are inexpensive so-called beginners' telescopes purchased as gifts by parents or spouses or by people with a bubbling enthusiasm for astronomy but little knowledge. My first telescope, bought under exactly those circumstances, was an inexpensive Japanese 60mm (2.4-inch) refractor. (The figure refers to the diameter of the main lens.)

That little instrument on its jiggly tripod gave me hundreds of hours of pleasure – and frustration. Pleasure because, for the first time, I could see things that I had read about in astronomy books. Frustration because, in the slightest breeze, the telescope would quake and shiver, making the object in view resemble the jumping dot of an oscilloscope.

More than anything else, that telescope taught me what to look for in the next telescope I purchased. First and foremost, I wanted a rock-steady mount. Second, I wanted larger and better optics that would give brighter, sharper images. However, I also knew from using the small telescope not to pay any attention to the magnification claims made by the manufacturers. A 60mm refractor – the most common type available then and now – cannot be used to advantage at magnifications exceeding about 120x. It is the same principle as a car speedometer that registers a maximum of 60 miles per hour faster than the car can possibly go.

But I learned all that *after* the purchase of my first telescope. And that problem persists today. Inadequate, poorly designed beginners' telescopes are still being sold to uninformed but well-intentioned consumers. Here are a few things to avoid when purchasing a telescope:

The classic novice telescope is most easily identified by its spindly tripod, either full-sized or (worse) tabletop format. The mounts have nice silver knobs and impressive dials, and the accessory boxes are full of (largely useless) gadgets. The instruments themselves are most often 50mm or 60mm refractors or

This rugged 15-year-old 60mm refractor is breaking in its third owner, and it still functions like new. Small refractors are ideal beginners' telescopes provided they have sturdy mountings and heavy-duty tripods. **Facing page:** *Equatorial mounts, which allow a telescope to compensate for the Earth's rotation and track celestial objects, can be set up in seconds by aligning the polar axis with Polaris. For general observing, hairline accuracy is not necessary.*

75mm to 100mm (3-to-4-inch) reflectors. Priced in the $75 to $300 range, these telescopes are available everywhere: department stores, camera shops, hobby-supply outlets. They may have brand names like Roddenstein, to give the aura of European optics, but virtually all such telescopes are made in Japan. I cannot recommend any of these telescopes, no matter whose name is on them.

The Japanese are unsurpassed in many fields, but the manufacture of these beginners' telescopes is not one of them. It is not necessarily poor optics that are at fault in these telescopes; it is everything else: the eyepieces, mounting, tripod, locking screws, slow-motion controls, finderscope and instruction booklet, which all range from poor to abysmal. Some beginners' telescopes are so inadequately designed that they are impossible to use for viewing anything but the moon. The cheapest ones, with plastic lenses, cannot be used even for that.

If these telescopes are so bad, why are they offered for sale? Where there is a market, there will be a product to fill it. People are attracted to a $200 telescope that looks as if it comes with an array of impressive accessories. Forget it. There is no such thing as a good beginners' telescope for under $300.

Good-quality amateur telescopes begin around the $400 mark and go up from there. Both U.S. and Japanese manufacturers compete in this realm. Indeed, most American brand-name telescopes have components made in Japan through contract arrangements. Many instruments with major American brand names like Celestron, Meade, Edmund and Tele Vue have anywhere from 10 to 90 percent Japanese components. It is a nice arrangement because the consumer often benefits from the best of both nations. That is not to say there are not first-class all-American telescope manufacturers. Two in this category are Astro-Physics and Questar. To balance the ledger, I will add Takahashi and Carton, two Japanese manufacturers that offer top quality in the over-$500 range.

The point is not American versus Japanese telescopes; it is quality versus junk. Avoid the lure of the department-store telescope. Ignore everything under $300. Price is the arbiter. If you can afford to go significantly over $500, you will greatly reduce your chances of being stuck with something you wish you had not bought. Plan ahead, and do it right the first time.

There are many advantages to bypassing the classic beginners' telescope, not the least of which is that a better, larger telescope is actually easier to use than the average small department-store telescope. The greater weight and stability of the larger instrument's mounting means that once the telescope is pointed at an object, it stays on target. Most of these larger telescopes are equipped with electric drives to track the celestial object across the sky.

All natural sky objects partake of an apparent motion because of the Earth's rotation. Electric motors integrated into the telescope's mount are geared to compensate for this motion. Once the polar axis of the telescope is aligned close to the star Polaris, the celestial target can be observed without having to make constant adjustments to keep it centred in the field of view. This is a real advantage, since the magnifications required for astronomy also magnify any tremor imparted to the telescope while trying to make such adjustments.

Finding sky objects is often much easier with bigger telescopes because most of them are equipped with adequate finderscopes — miniature telescopes mounted parallel to the main tube that allow easy alignment on the target object. Once centred in the cross hairs of the finderscope, a celestial body is automatically centred in the main telescope. Finderscopes have much wider views of the sky than the main instrument, making it easier to zero in on the desired object. In my experience, finderscopes on less expensive telescopes usually have inadequate optics and mounting brackets barely stronger than paper clips, which make it difficult to align the finderscope with the main telescope. Something as fundamental as locating a certain sky object then becomes a frustrating experience. "The larger, the better" is the maxim for finderscopes. The 5 x 24 size is inadequate; 6 x 30 barely acceptable; 8 x 40 and 8 x 50 far superior. An alternative to finderscope upgrading is the Telrad ($50), an illuminated sighting device easily added to any telescope.

Don't make the same mistake I did by buying an inferior beginners' telescope. Be patient, learn your way around the sky with binoculars, and save your money for a larger good-quality telescope.

TELESCOPE TYPES

There are three main types of telescope optical systems used in backyard astronomy: refractors, Newtonian reflectors and Schmidt-Cassegrains. **Refractors** of all sizes are basically sophisticated spyglasses that have a main lens concentrating the incoming light to a focus at the lower end of the tube, where a magnifying ocular called an eyepiece provides a visual image. Refractors with a main lens between 60mm and 100mm (2.4 to 4 inches) are excellent performers, rugged and ideal for novice backyard astronomers. Unfortunately, as I mentioned earlier, 60mm refractors outfitted with mediocre trappings, marketed in camera and department stores, have given refractors in general an undeserved bad reputation among serious amateur astronomers, who often consider them unworthy. Many excellent refractors are available. Celestron, Meade and Tele Vue, for example, offer 75mm-to-100mm refractors with quality mountings, eyepieces and other furnishings, along with a first-class lens. Prices range from $500 to $2,000.

Theoretically, refractors are the highest-performance telescopes for backyard skywatchers, but for years, their prices put them out of contention in apertures where they perform best: 4 to 6 inches (100mm to 150mm). If price did not deter the observer, the awkward four-to-seven-foot tubes would. However, recent innovations in optical glass and lens design have shrunk the tube length while retaining, or even improving on, the refractor's superb performance. These instruments, called apochromatic (which means colour-free) refractors, offered by Astro-Physics, Meade, Celestron, Tele Vue and others, have introduced the refractor to serious backyard astronomy. A 4-inch apochromatic refractor, with equatorial mount, sells in the $2,000 range.

Refractors over 4 inches were almost

TRASH SCOPE BLUES

"Am I doing something wrong, or is it this telescope?" the voice on the telephone was asking. "It doesn't seem to focus, and I am never sure what I am pointing at."

I was speaking with another frustrated owner of a $200 department-store telescope, the kind that comes in a package announcing "450-Power Astronomical Telescope." The caller admitted that the impressive complement of accessories included with the instrument as well as the packaging embellished with colour photographs of comets and nebulas had been too much to resist. But now, he was wondering whether he had made a mistake.

"I can't seem to keep it steady — everything's a blur," he moaned. "The only eyepiece that seems to show anything at all is the one marked K20mm. What am I doing wrong?"

"Nothing," I sighed. I had heard it all before — hundreds of times. "It's not you, it's the telescope," I assured him.

I went on to explain that his department-store 60mm refractor telescope is an example of what many experienced amateur astronomers uncharitably, but aptly, call "Christmas trash scopes." I told him that his frustration was perfectly normal. The classic trash scope is designed not for ease of use but to appeal to well-meaning but uninformed gift buyers and rank beginners. Well-known brand names also lure the novice, but they mean nothing, since all telescopes in this class are made in the same giant factories in the Orient regardless of whose name is on them.

Even expert observers soon become frustrated trying to operate these instruments, with their jiggly mounts and rickety tripods. They barely function, usually only at their lowest power, typically about 36x with a 20mm eyepiece. Accessories such as Barlow lenses, image erectors, sun-projection screens, filters and high-power eyepieces (5mm to 12mm) supplied with these telescopes are so cheaply made that they are almost impossible to use and are included to give the impression of a fully equipped instrument.

If you have a telescope like this, use it with the 15mm-to-25mm low-power eyepiece and the right-angle prism diagonal for introductory views of the moon, Saturn's rings, Jupiter's moons and a few double stars. Don't expect much more. And if you do not already own a telescope like this, consider yourself lucky to have read this first.

nonexistent as amateur equipment until recently because their tubes were too long, they were too expensive and they were afflicted with chromatic aberration, which produces unwanted colour haloes around bright objects. Now shorter, less expensive 5- and 6-inch apochromatic refractors pioneered by Astro-Physics in the 1980s are available with performance characteristics that equal significantly larger-aperture telescopes of other optical configurations. Nonetheless, refractors are still the most expensive type of telescope for a given aperture (with the exception of the Questar telescope, mentioned later).

Until the mid-1970s, the telescope that dominated amateur astronomy for more than two decades was the **Newtonian reflector**, which uses a precision-ground shallow bowl-shaped mirror at the base of an open tube to reflect light and bring it to a focus near the top of the tube. A smaller flat-surface mirror angled at 45 degrees and suspended at the top of the tube reflects the light out a hole in the side, where the focuser is located. With the focuser at the top of the tube, a Newtonian offers a comfortable eyepiece position for overhead viewing, while other types are at their worst looking straight up. However, large Newtonians require ladders. Because they are easier to manufacture than refractors, Newtonian reflectors can be made in larger sizes, at a cost within reach of any serious backyard astronomer. They are marketed in sizes from 3 inches up.

As with refractors, the most inexpensive Newtonian reflectors should be avoided. In the 3- and 4-inch size range, they are often equipped with inferior mountings and accessories. Another danger which the novice will likely be unaware of, even after purchasing the telescope, is that the mirror system can be jarred out of alignment during transport or with daily use. Realigning the mirrors is not difficult, but the inexperienced observer who sees fuzzy, cometlike images of stars usually blames the optics. Alignment problems aside, Newtonian telescopes offer good value for the money.

Newtonians range in price from $50 to $200 per inch of aperture, depending on quality of optics and the sophistication of the mount. Many veteran observers who are serious about their observing own Newtonians. A well-made, well-maintained Newtonian can provide outstanding performance.

Newtonian reflectors come on two types of mountings – equatorial and Dobsonian. When properly aligned, the equatorial mount will track the stars. Equatorial mountings are functional and manageable for 6-to-8-inch Newtonians, but they become cumbersome for 10-inch telescopes and are downright massive in larger sizes. Transporting one of these behemoths to and from a dark observing site is a major expedition. (The ritual of spending an hour dismantling the telescope at 2 a.m. is enough to discourage all but the most rabid enthusiast.) This is why Dobsonian mounts invaded the amateur-astronomy scene in the late 1970s, when Newtonians 10 inches in aperture and greater were becoming increasingly popular.

Dobsonian mounts are simple vertical-horizontal affairs known as altazimuth (altitude-azimuth) mounts, but their simplicity translates into lighter, more compact mounts with surprising stability. The price is also attractive: about $500 for a 10-inch model. Named after California amateur astronomer John Dobson, who popularized the design, the Dobsonian has become the preferred mount for large Newtonian reflectors. The Dobsonian's inability to track a celestial target automatically (it has to be manually recentred about once a minute) is the price that must be paid in exchange for portability. Amateur astronomers are now mounting telescopes up to 24 inches in aperture on these simplified pedestals. Such large telescopes are sometimes referred to as "light buckets" because they collect far more starlight than their smaller cousins. However, they are strictly deep-sky instruments suitable for galaxies, nebulas and star clusters; they actually perform worse than smaller telescopes for viewing the moon, planets and multiple stars. But one look at a globular cluster or galaxy through a light bucket, and it is hard to deny the lure of these large telescopes.

In addition to the introduction of Dobsonian mounts, the 1970s saw an even more important revolution in amateur telescopes: the mass production of **Schmidt-Cassegrain** systems. A Schmidt-Cassegrain telescope combines many of the best features of the refractor and the Newtonian reflector. It has a concave main mirror like the Newtonian, but a lens at the top of the tube performs the triple function of correcting for optical aberrations, sealing the tube from dust and other airborne pollutants and supporting a second mirror that reflects the concentrated light back through a hole in the main mirror. The light is finally focused at the rear of the telescope.

Although these instruments produce slightly less crisp images than top-of-the-line Newtonians or a somewhat smaller refractor, their main advantage is that the tube is shrunk to an extremely compact configuration. An 8-inch Schmidt-Cassegrain is about two feet long, compared with five feet for a typical 8-inch Newtonian. More telescope is compressed into a smaller package, and that is the main reason why Schmidt-Cassegrains have emerged as the most popular type for serious backyard astronomers. Celestron and Meade are the big-two American manufacturers of Schmidt-Cassegrain telescopes, and both offer excellent value for the money. Eight-inch Schmidt-Cassegrains are priced from $1,500 to $4,000. Schmidt-Cassegrains therefore cost somewhat more than Newtonians of comparable aperture, especially if the Newtonian is outfitted with a Dobsonian mount, so there are some trade-offs between the two types. But for portability – and this is a major consideration – the Schmidt-Cassegrain can't be beaten.

A variation of the Schmidt-Cassegrain form, which actually predates it as amateur equipment, is the Maksutov-Cassegrain. The Maksutov-Cassegrain is most familiar as the Questar telescope, an incredibly portable 3½-inch telescope that comes in its own fitted leather case. Although they are the most expensive instruments in their aperture range (about $4,000), Questar Maksutov-Cassegrains are widely praised by observers for their razor-sharp imagery.

MAKING THE CHOICE

What type and size of telescope is the best all-purpose buy? It would be nice if there were a definitive answer to this question, but because of the many variations available, there is not. Each has advantages and disadvantages. I have owned and used dozens of instruments, ranging from small refractors to giant Newtonian reflectors. Personally, I prefer telescopes in the moderate-aperture range – 4 to 8 inches. Their ease of use, portability and ability to show all types of celestial objects make them the workhorses of amateur astronomy.

I have owned, and since disposed of, several larger instruments, including those I call the hernia models (which are convenient to set up only when two people are available) or the Marquis de Sade models (which come in several bulky pieces and have a plethora of nuts and bolts and clamps to be manipulated, as well as sharp corners to scratch elbows and shins). Lack of portability has relegated many such unwieldy models to enthusiasts' storage closets. Aperture fever is a common affliction among amateur astronomers. But until one actually handles telescopes, it is difficult to realize how much bulkier instruments become in 10-inch-and-larger sizes. Once the initial euphoria wears off, it becomes extremely tiresome to lug out one of those cumbersome monsters. Even some of the stripped-down Newtonian

Inexpensive and portable, the mount and tube for this 8-inch Dobsonian-mounted Newtonian reflector, **above**, *was entirely homemade. David Levy,* **left**, *discoverer of several comets, uses a 16-inch Newtonian reflector. Newtonians have open tubes with the main mirror at the lower end. Supported near the top of the tube, a smaller mirror at a 45-degree angle deflects the converging light from the main mirror to the side-mounted focuser. The author,* **far left**, *with a 10-inch Schmidt-Cassegrain.*

telescopes on Dobsonian mounts are still hefty pieces of equipment.

Apart from lack of portability, there is another reason why monster telescopes may not perform to expectations. Large optics of excellent quality must be made by hand, an expensive and time-consuming procedure. A big, inexpensive telescope cannot possibly have high-precision optics; consequently, the images may be brighter but are often fuzzier too. Second, the larger the telescope, the more likely it is to be affected by a phenomenon known as poor seeing. *Seeing* refers to the steadiness/unsteadiness of the image of a celestial object viewed through the telescope. (Good seeing = steady image; poor seeing = unsteady image.) Turbulence in the Earth's atmosphere imparts a shimmering quality to telescopic images. The intensity of the turbulence varies depending on winds, temperature differential among upper-atmospheric layers, and air circulation immediately around the telescope. The larger the telescope, the worse it is affected, since large telescopes have to peer through more air than smaller ones because of the differing surface areas of their optics. For example, a telescope with a main mirror or lens 8 inches in diameter has to look through a column of air 8 inches wide and about 10 miles long, the nominal thickness of the turbulent layers of the Earth's atmosphere.

So the common notion that professional astronomers using the largest observatory telescopes automatically have the best views of the cosmos is generally false. Observatory telescopes may be 10 times larger than backyard telescopes, but the view is *never* 10 times better. Sometimes it is no improvement at all — brighter images, but no more detail. The huge telescopes used for frontier research are not designed for visual use anyway. Sophisticated electronic, photographic and spectroscopic detectors, rather than the human eye, are placed at the focus. For those detectors to function, the telescopes must produce a bright image; therefore, the instrument must be as big as possible.

Large telescopes are also plagued by "cool-down time" — the interval, usually shortly after sundown, during which the optics are radiating heat and their optical figure is slightly distorted. Telescopic images then have a boiling appearance. The cool-down factor is even more prominent when a telescope is taken from the house into the colder air outdoors. Some instruments require hours to stabilize.

However, a large telescope will always produce a brighter image. Light-collecting ability varies with the square of the aperture. A 12-inch telescope produces images nine times brighter than a 4-inch. Therefore, objects nine times fainter can be seen. For example, a 4-inch telescope will reach 13th magnitude, an 8-inch 14th, and a 12-inch will probe to 15th magnitude. This advantage is best applied to faint objects like nebulas and galaxies, which show up far better in larger telescopes. But even in these cases, the effects of poor seeing take a toll. Furthermore, a black sky is essential for such deep-sky observing, no matter what instrument is used. If the telescope has to be transported to a dark site, its size can become a liability.

All of these factors explain why the most popular telescopes are in the 4-to-8-inch range, the ideal size for portability and the best compromise between large telescopes, which are somewhat handicapped by sheer bulk and the effects of poor seeing, and small instruments, whose images are sharp but dim.

What is the bottom line? When pressured to recommend a first telescope, I usually suggest a solidly mounted Schmidt-Cassegrain in the 4-to-8-inch range, a

An observatory protects the observer from wind, dew and nearby lights. Portable model, **above left,** *was constructed using boating supplies. Permanent observatory,* **above right,** *also homemade, houses an 8-inch Schmidt-Cassegrain telescope.* **Far right,** *three serious backyard telescopes: 10-inch equatorial-mounted Schmidt-Cassegrain, 5-inch equatorial-mounted refractor and 13-inch Newtonian reflector on Dobsonian mounting. These telescopes are close to the maximum convenient size for portability in their respective classes.*

"HOW POWERFUL IS IT?"

Whenever I demonstrate one of my telescopes, I am invariably asked, "How much did it cost?" and "How powerful is it?" For anyone contemplating a telescope purchase, the cost is certainly an important consideration. But the question about power focuses on one of the most misunderstood aspects of telescope performance. Claims of huge power capabilities are almost totally meaningless.

There are three distinct types of telescopic power — light-gathering, resolving and magnifying. Least important is magnification, yet magnifying power alone is often used as a selling point for small telescopes. The most significant ingredient is light-gathering ability. For example, the Orion Nebula will look about the same size in a 2-inch telescope as as it does in a 4-inch if each is used at the same magnification. However, the light-gathering power of the 4-inch telescope is four times greater than the 2-inch, so the nebula will be four times brighter.

This is an important difference, since most astronomical objects are relatively faint and need their brightness significantly boosted for proper examination. The images must be bright before they can be subjected to magnification. Only the sun, moon and brighter planets are sufficiently luminous so that light-gathering power is not a crucial ingredient for a decent view.

Given adequate light-gathering ability, there are other limits on magnification. As a general rule, the practical maximum magnification limit is about 50x for each inch of aperture. This means that the upper limit for a standard 60mm (2.4 inch) refractor is 120x. When such a telescope is pushed to 200x, the images are grossly overamplified and exceed the third factor on our list: resolving power, the ability of the instrument to discriminate fine detail. The limitation on resolving power is imposed by the interaction of light and optics. Even the best optics exceed their resolving-power limit above 50 magnifications per inch.

Excessive magnification not only exceeds the telescope's capabilities by producing grossly fuzzed images but also makes the instrument almost impossible to use. Tiny jiggles created when the focusing knob is touched or movements generated by a breath of wind become amplified by the same amount that the instrument is magnifying, causing the star or planet to quiver or lurch across the field of view. A further drawback is the tiny eye-lens opening in the eyepieces required to reach high powers. Sometimes, it is like trying to peer through a pinhole. The narrow field of view means that more time must be spent locating and centring sky objects. And, once located, celestial objects soon drift out of the field of view due to the Earth's rotation (unless the telescope is equipped with an equatorial mount and drive).

And if all this has not discouraged the power-hungry tyro, then the ubiquitous problem of atmospheric turbulence and its frequent poor seeing conditions will be the final damper. The 50x-per-inch limit only applies in good seeing, when air turbulence is minimal. Telescopes are frequently seeing-limited to 20x to 30x per aperture inch.

Normal operating magnification for astronomy is 4x to 30x per aperture inch. This range offers the best ratio between aperture and magnification by providing the correct balance of light-collecting and magnifying power. On my 6-inch telescope, for example, I rarely use magnifications over 180x, and 40x provides the most stunning views of brighter star clusters.

If telescope advertising trumpets power capabilities beyond the limits of even the best possible optical systems, look elsewhere.

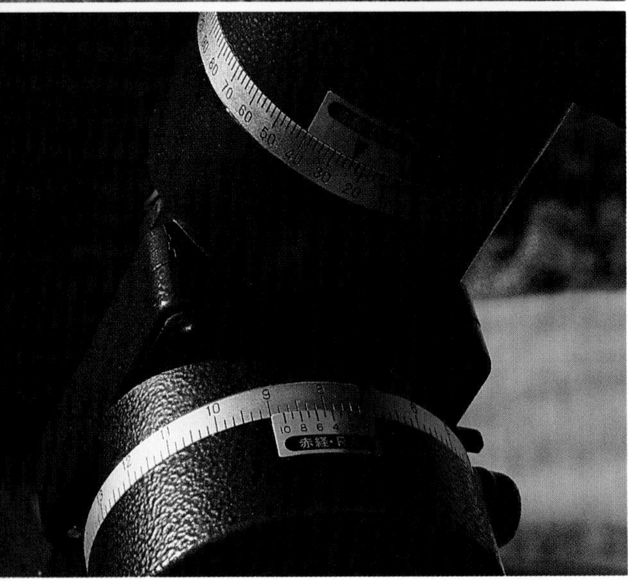

Amateur astronomer Steve Dodson, **top,** *poses beside his colossal 22-inch trailer-mounted Newtonian. Setting circles,* **above,** *can be used to dial in desired celestial objects, but many observers prefer a direct visual hunt. Telescope eyepieces,* **right,** *are available in three barrel diameters: 0.965-inch, 1¼-inch and 2-inch.* **Far right:** *An array of eyepieces, Barlows, binocular tripod adapters and other accessories.*

6-inch equatorially mounted Newtonian reflector, an 8- or 10-inch Dobsonian-mounted Newtonian or a 3-to-4-inch refractor. While it seems incongruous that such small refractors should even be considered along with these larger instruments, I cannot dismiss them. I had many wonderful moments under the stars with my 3-inch refractor and have always been pleased by the superb imagery of my 4-inch, which is my most-used telescope. Secondhand refractors are an excellent buy because they virtually never wear out or need repairs. If well maintained, all of these instruments have a resale value of at least 60 percent of their original price 5 to 10 years after purchase.

So far, I have not mentioned building a telescope. A generation ago, commercial telescopes were a luxury item, substantially more expensive in terms of personal income than is the case now. Active participation in amateur astronomy usually meant becoming a telescope maker. Today, the vast majority of amateur astronomers purchase commercial instruments and leave the telescope making to a small band of devotees. Making your own is no longer a way to save money. Telescope making requires many months of part-time labour grinding mirrors or lenses, fabricating tubes and mounts and generally reinventing the wheel. I would sooner spend my time under the stars.

TELESCOPE ACCESSORIES

Most commercial telescopes have long lists of optional accessories. In some cases, even the tripod is listed as "optional." Amateur astronomers usually buy a minimum of options initially, adding items as required. This is a wise policy. It allows the owner to get used to the equipment, recognize its deficiencies and evaluate the need for various accessories. Here is a rundown of the telescope accessories most often requested by new telescope owners.

Solar filters: Priced from $50 to $200, these highly desirable aids attach to the front of the telescope to reduce solar radiation to suitable levels for direct viewing (see Chapter 8).

Slow-motion controls: Most equatorially mounted telescopes offer these as standard equipment, permitting easy centring of celestial objects in the eyepiece field of view. Generally a useful aid.

Telecompressors: These are Schmidt-Cassegrain accessories for low-power deep-sky sweeping. The lenses intercept the light before it enters the eyepiece, widening the field of view ($80 to $250).

Erecting prisms: Sometimes called porro prisms, they fit into the focuser ahead of the eyepiece for right-side-up land viewing. (All astronomical telescopes invert the image or reverse it left to right.) These are used less often than might be assumed when the telescope is first purchased. Usually only practical on small refractors.

Nebula filters: The glow from cities, shopping centres and streetlights dims the natural beauty of the night sky. A nebula filter attached to the telescope's eyepiece blocks much of the interfering glow, producing better contrast in views of nebulas. Priced from about $60 each, they work best on short-focal-ratio telescopes. Nebula filters provide minimal advantage on objects other than nebulas.

Dew caps: Dew forming on the front lens plagues owners of Schmidt-Cassegrain and Maksutov-Cassegrain telescopes. An extension of the telescope tube, known as a dew cap, tends to prevent dew formation. Another solution is to use a hand-held hair dryer to evaporate the dew.

Photographic accessories are definitely best left until you have had some experience with the telescope in straightforward visual observing. Avoid the temptation to load up on adapters, off-axis guiders and dual-axis slow-motion motors until you are fully familiar with the comparative difficulties involved in various types of astrophotography. Most of the types of photography suggested in Chapter 11 require no adapters and should be tried first before plunging into more difficult aspects of celestial portraiture.

Telrad is a trademark for a unique one-power bull's-eye sighting device that makes a dandy finderscope.

Motorized focusers are useful on telescopes that quiver when you touch the focus knob — which means most telescopes. However, this accessory is highly personal. Some observers can't be without push-button focusing; others prefer the manual method.

Computer controls that point your telescope toward any one of thousands of celestial objects in the computer's memory banks when you simply key in the object's name sound like just the thing for a beginner learning his or her way around the sky. The trouble is, you will never learn anything if the computer is doing all the finding. For more on this, see "Dial-A-Galaxy" (pages 81-82).

EYEPIECES

Sometimes, telescopes with otherwise excellent optics are supplied with bottom-of-the-line eyepieces. A good optical system will perform to capacity only when linked to top-quality eyepieces. Inexpensive beginners' telescopes are usually equipped with Japanese standard-sized eyepieces — 0.965 inches outside diameter (the width of the section that slips into the eyepiece holder on the telescope). The American standard is 1.25 inches.

In recent years, manufacturers have responded to a desire among amateur astronomers for high-quality eyepieces by developing new and significantly better specifications in the 1.25-inch size. If a telescope has a 0.965-inch-diameter eyepiece, the only solution for better performance is to change the focusing mechanism to the higher diameter or to buy a 1.25-inch adapter or a hybrid diagonal (0.965 inches on one end, 1.25 inches on the eyepiece end). Even some fairly high-priced telescopes are not supplied with top-of-the-line eyepieces. Here is what you need to know to make informed choices.

Just as there are different types of telescopes with differing optical configurations, eyepieces (sometimes called oculars) come in a range of types, each with its own advantages and disadvantages:

Huygenian and *Ramsden*, the simplest eyepiece types available, have been around for centuries. They accompany only the cheapest telescopes, since they are worth just a few dollars each. They have been completely superseded by the other designs and are usually found in 0.965-inch sizes. They can be identified by the letters R, H or AR engraved on the outside. Anyone with an inexpensive telescope undoubtedly has one or two of these.

I also rank *zoom* eyepieces in the undesirable category for astronomical use. They seem like a good idea, but in practice, the field of view is restricted and the optical performance (in all the zoom eyepieces I have ever used) is fair to poor.

The *Kellner* eyepiece is also an older design, dating back to the 19th century, but it is still in use today on refractors and Schmidt-Cassegrains and provides acceptable performance in low- or medium-power applications. Modest price ($30 to $60) is its prime attribute. Its main disadvantage is a relatively narrow field of view. An improvement on the Kellner design, which reverses the arrangement of the three lenses in the eyepiece and utilizes modern high-index glass, is the RKE series of eyepieces, developed by Edmund Scientific in 1978. Other manufacturers followed suit with similar designs under different names, all of which yield better imagery and a wider field than the standard Kellner but still at a relatively low price, usually less than $50.

Orthoscopic eyepieces were once regarded as the finest available, and for many years, they were. They improve on the Kellner design by adding a fourth lens element, which provides a wider field of view and eliminates virtually all optical aberrations. Price range is $55 to $85. Orthos are excellent eyepieces, preferred by many backyard astronomers for medium- and high-power applications.

Erfle eyepieces became popular after World War II because the five-element design was widely used in military optics. They are favoured for low-power applications because of their extremely wide field of view, almost twice that of a Kellner and substantially wider than an orthoscopic.

A variation on the Erfle, known as the *König* design, has generally sharper definition with a slightly narrower field of view, although both types produce mild "ghost" images on bright objects. Königs are superb low- and medium-power eyepieces priced in the $60 to $150 range.

Plössl eyepieces are in many ways superior to the types mentioned so far. Their four-element design provides remarkably sharp images with a slightly wider field of view than the orthos. Plössls are extremely versatile, being suited to low-, medium-

and high-power applications. Prices range from $60 to $120 each. I prefer Plössls for most applications, although I still like Erfles and Königs at lowest power. For observers on a budget, Kellners and the RKE family are the best buys.

Wide-field eyepiece designs with six or seven elements are a recent advance on the Erfle design and offer better overall imagery. They are very expensive, especially in lowest-power versions that require a lot of glass. A new wide-field design, *Panoptic* by Tele Vue, is the premium entry in this class, offering superb low-power definition.

Another elegant eyepiece design is the *Nagler* series introduced in 1981 by Tele Vue, undoubtedly the finest medium- and high-power eyepieces available. A similar design, called *Ultra Wide Angle*, was introduced by Meade a few years later. The eyepieces cost as much as a small telescope ($200 and up), but their razor-sharp, extremely wide fields of view noticeably improve the performance of any telescope. However, all that glass means these are heavy eyepieces which can overweigh a small or delicately balanced telescope.

The magnification provided by an eyepiece can be determined by dividing the focal length of the telescope by the focal length of the eyepiece. A 2000mm focal-length telescope using a 25mm eyepiece will yield $2{,}000 \div 25 = 80$x. The eyepiece focal length is always engraved on the side, along with the type or a letter indicating the type (e.g., K25mm = Kellner 25mm focal length). Low-power eyepieces range from 40mm to 20mm focal length; medium power from 19mm to 13mm; high power from 12mm to 4mm. I recommend having at least one of each or a low and medium eyepiece and a *Barlow*.

A Barlow is basically a tube with a lens at the lower end and fittings to accept an eyepiece at the upper end. A Barlow amplifies the resulting magnification of the eyepiece two to three times, depending on the manufacturer's specifications. I almost never use eyepieces shorter than 12mm focal length. A 2x Barlow and a 16mm eyepiece are much more comfortable to use than an 8mm eyepiece alone. Both yield exactly the same magnification. A typical Barlow costs about the same as a good-quality eyepiece. Beware of inferior-quality Barlows supplied as standard equipment on some telescopes.

Many larger telescopes can accommodate 2-inch-diameter eyepieces, which, although they can cost more than $200, can provide stunning low-power performance with fields of view substantially wider than any 1.25-inch eyepiece.

Eyepieces are important. Quality eyepieces improve any telescope's performance. Your eyepiece collection should be worth at least one-third of the value of the telescope. Eyepieces have excellent resale value and can be used on any telescope, so your collection has permanent value.

With hundreds of eyepieces now available, the decision about what to buy is tougher than ever. I have used many of them, and here are some of my favourites: 40mm and 32mm Königs (2-inch size) by University; 35mm Panoptic by Tele Vue; 30mm to 15mm Plössls by Celestron, Meade and Tele Vue; 21mm RKE by Edmund; 18mm Super Wide by Meade; 16mm König by University; 16mm Nagler by Tele Vue; 9mm and 7mm Naglers by Tele Vue or 8.8mm and 6.7mm Ultra Wides (Nagler type) by Meade. I also like the 2.5x Tele Vue Barlow. You will never need all of these, and there are many other fine eyepieces besides those mentioned.

Although I wear glasses, I seldom use them when at the eyepiece. A slight change in focus accommodates all vision differences except astigmatism. (Those who wear glasses to correct astigmatism should leave them on.) Most eyepieces, especially medium- and high-power, require the eye to be closer to the eye lens than eyeglasses permit. Keeping glasses on reduces the field of view but otherwise does not substantially alter the imagery.

TELESCOPE FOCAL RATIO

Focal ratio is something you will need to know about both before and after you become a telescope owner. Every telescope operates at a specific focal ratio. The focal ratio is usually marked on the instrument or supplied in the owner's manual. Examples are f/10 or f/4.5. The focal ratio is determined by dividing the telescope's focal length (the distance that the main lens or mirror refracts or reflects light to the point of focus) by the diameter of the main lens or mirror. Thus an 80mm refractor with a 1200mm focal length is f/15, and a 10-inch Newtonian of 45-inch focal length is f/4.5.

Newtonian reflectors and refractors are about as long as their focal lengths, so just measuring the length of the tube and dividing by the aperture gives a rough focal ratio. Not so with catadioptrics. These systems fold the light path and squeeze a long focal length into a short tube. Schmidt-Cassegrains are usually f/10, although some newer models are f/6. Maksutov-Cassegrains range from f/7 to f/16. Newtonians vary from f/4.5 to f/10, and refractors are f/12 to f/16 (standard, or achromatic, versions) or f/6 to f/9 (apochromatic). For a variety of reasons, the smaller the focal ratio, the more difficult it is to manufacture good optics.

One persistent myth, perpetuated by some telescope manufacturers, is that telescopes with small focal ratios produce brighter images than telescopes with large focal ratios. This is true *only* for photography through the instrument. Visually, any telescope of a given type and aperture (say, two 8-inch Schmidt-Cassegrains) will produce identically bright images *when used at the same magnification*, even though they have different focal ratios. What is different is that telescopes with smaller focal ratios are capable of operating at lower minimum power. However, in most instances, f/6 is short enough for all visual applications. There is no ideal focal ratio, although I often have found the most versatile telescopes of all types are f/6 to f/8.

Binoculars and telescopes are the amateur astronomer's passport to the universe. Selecting the right equipment requires consideration of many factors outlined in this chapter.

TELESCOPE COMPARISONS

Type	Variations	Aperture Range in Common Use	General Specifications/Performance	Applications
Refractor	achromatic	2.4" to 4"	These rugged, reliable, generally maintenance-free instruments have been a traditional "first" telescope for backyard astronomers. Easily portable in 3-inch and smaller sizes. Beware the inadequate mounts and spindly tripods on less expensive models.	Excellent for moon, planets, star clusters, double stars and general introductory scanning. Poor on nebulas, faint clusters and galaxies due to small apertures. Best buy for city or suburban sky viewing where faint objects are obscured anyway.
	apochromatic	3" to 7"	Unquestionably the finest telescope optical systems available because of totally aberration- and obstruction-free lens design. Superb imagery. Expensive. Easily portable in 4-inch size.	Outstanding lunar and planetary performance in all sizes. Ultrasharp star images and good deep-sky penetration. Only limited by relatively small aperture, compared with Newtonians.
Newtonian Reflector	equatorially mounted	4" to 12"	Least expensive of all types in comparable aperture. Requires more maintenance than other types. Cumbersome in sizes over 8-inch. If well made, can provide excellent value and fine performance, especially in 6-inch size.	Yields good results on all types of backyard telescopic activities; especially appropriate well away from light-fogged skies.
	Dobsonian mounted	8" and up	Simplified lightweight mount and short focal ratios result in large aperture in a relatively compact and inexpensive package. Very portable in 8-inch size.	Must be used under dark-sky conditions. Excellent for nebulas, clusters and galaxies. Poor on moon, planets, double stars. Tracking must be done manually at all times due to simplified mount design.
Compound	Schmidt-Cassegrain	4" to 14"	Optical performance generally good. Stubby tube results in stable, easy-to-use system. More expensive than Newtonian types, but only modestly so.	Compact design and vast array of accessories for all applications. Widely regarded as the best all-round instrument for backyard astronomy. Astrophotographers' favourite.
	Maksutov-Cassegrain	3.5" to 7"	Well known in Questar design for excellent performance and compact format but has never dominated the amateur-astronomy scene. Relatively expensive except in "spotting scope" designs, which are not recommended for astronomy.	Good for every backyard application except wide-field low-power viewing. Small models are extremely compact, excellent for frequent travellers.

FACTORS TO CONSIDER WHEN SELECTING A FIRST TELESCOPE

Telescope type	Will telescope usually be transported to observing site?	Is main observing site an urban location or a dark rural location?	Will telescope be used for celestial photography?
2.4" to 3.5" refractor	Generally easily transportable.	Good performance in urban environment; will not equal other types in dark rural skies.	Not recommended except for most basic shooting.
4" to 7" apochromatic refractor	Easily transported in 4" size; 6" and 7" are hefty.	Best bet if you are often limited to urban observing.	Becoming a favourite for astrophotographers.
6" to 10" Newtonian reflector, equatorial mount	Generally transportable, but may be a tight fit in a small car; 10" models can be brutes.	Performance limited in urban environment; excellent performer in dark skies.	Will yield excellent results with proper accessories.
10" or larger Newtonian, Dobsonian mount	Transportable, but a station wagon or van may be needed.	Urban environment severely limits usefulness of this type; designed for use in dark skies.	Not recommended.
4" to 8" Schmidt-Cassegrain	Easily transportable in any vehicle.	Good all-round performer.	Excellent range of accessories makes this the most frequent choice of backyard astrophotographers.
3.5" to 7" Maksutov-Cassegrain	Exceptionally compact in smaller apertures.	Often a favourite for urban astronomers who occasionally observe from really dark sites.	Long focal ratio* has some disadvantages, but performance is generally good.

Telescope type	Will telescope be used primarily for viewing sun, moon, planets and other bright, easy-to-find objects?	Will telescope be used primarily for viewing nebulas, star clusters and galaxies?	Will telescope frequently be used for daytime land viewing?
2.4" to 3.5" refractor	Excellent performance due to clean images from unobstructed light path. Long focal ratios* and equatorial mount preferred.	Good performance in medium and short focal ratios*; small apertures give excellent portability for teaming up with larger telescopes for detailed views.	Suitable and recommended.
4" to 7" apochromatic refractor	Unsurpassed for consistently fine lunar and planetary performance. The preferred choice if you can afford one.	Excellent for their aperture, but outperformed by larger Newtonians.	Tele Vue "Genesis" and Astro-Physics "Traveler" especially recommended for daytime viewing.
6" to 10" Newtonian reflector, equatorial mount	Often outstanding results obtained in medium focal ratios with good optics. Short-focal-ratio telescopes noticeably less sharp on these objects due to larger central obstruction.	In short and medium focal ratios, these telescopes are top-rated for deep-sky viewing. Wider fields of view than other types, although edges are degraded by coma.	Not recommended.
10" or larger Newtonian, Dobsonian mount	These telescopes are not designed for observing bright objects and seldom produce satisfactory views compared with other types.	These telescopes are becoming very popular as the most economical way to see faint objects. Images are often fuzzy at higher magnifications but excellent at low powers, which are used most of the time anyway.	Not recommended.
4" to 8" Schmidt-Cassegrain	Although these telescopes produce acceptable performance in this category, images always have a slightly gauzy appearance due to effects of large central obstruction.	Deep-sky performance rivals that of any other telescope type. Although more expensive than Newtonian designs, Schmidt-Cassegrains are more compact and have convenient eyepiece location.	Suitable in smaller sizes.
3.5" to 7" Maksutov-Cassegrain	When well made, this design seems to produce performance second only to refractors. However, top-quality Maksutov-Cassegrain telescopes are very expensive.	Since most Maksutov-Cassegrains are in the 3½-inch-aperture range, these instruments are less effective than other types simply because of low light-collecting power. Long focal ratio is sometimes a handicap as well.	Suitable in smaller sizes.

*Short-focal-ratio telescopes are usually f/4 to f/5; that is, the focal length (effective distance from main optical element to eyepiece) is four or five times the diameter of the objective lens or mirror. Medium-focal-ratio telescopes are f/6 to f/8; long-focal-ratio, f/9 to f/16.

Selecting a telescope requires careful consideration. How frequently the telescope will be used in well-illuminated urban or suburban environments is a key factor, since some instruments are designed specifically for dark-sky sites. For most observers, though, portability and quick set-up time are the most important considerations.

6

Probing the Depths

What is inconceivable about the universe is that it should be at all conceivable.

— Albert Einstein

An acquaintance who was aware of my interest in astronomy noticed my telescope in the backyard at dusk one evening. "What do you actually *do* with that?" he asked. I launched into an enthusiastic description of the universe of planets, stars and galaxies. After listening attentively, he then wondered what I did after I had seen all of that.

I tried to explain that I spend most of my time reexamining objects that I have seen before, but I realized how difficult such a concept must seem. Backyard astronomers are a special breed. They savour their moments under the stars. They have an infatuation, a love affair, with the cosmos that grows and nurtures itself just as meaningful human relationships do. Of course, it is a less definable, one-way relationship, but I have come to regard that feeling as the closest I can ever come to being at one with nature. After a night under the stars, I have a sense of mellowness, an amalgam of humility, wonder and discovery. The universe is beautiful, both visually and in a deeper sense.

The visual beauty is at least clearly definable. It comes in a variety of forms, ranging from stars with heartbeats to villages of suns perched in the Milky Way Galaxy's nearby spiral arms to remote stellar cities, the galaxies. It is now time to be specific, to define exactly what can be seen with a telescope or binoculars and where in the sky to look. In the following chapters, we will examine solar system phenomena — the sun, planets, their satellites and comets — and nighttime atmospheric phenomena — meteors and auroras. But all of that is our cosmic backyard. The universe beyond the solar system, explored in this chapter, is known to amateur astronomers as the deep sky, and targets of interest there are commonly called deep-sky objects.

Even though most stars in the nighttime sky are bigger and brighter than our sun, nothing but a tiny point is seen telescopically, no matter how large the instrument or what magnification is used to observe a star. Stars are simply too far away. Under high magnification, a star actually does appear as a disc, but that results from the nature of optics and from their interaction with light. It is called the Airy disc, after the 19th-century English astronomer who explained it.

DOUBLE STARS

Despite the fact that most stars are solitary dots when viewed through a telescope, single stars are in the minority. At least half, and possibly as many as 80 percent, of all the stars in the Milky Way Galaxy are members of double- or multiple-star systems — two or more suns gravitationally bound and orbiting about one another. Quite often, these stars orbit more closely than one astronomical unit (AU). To be visible in backyard telescopes, the components of a binary or multiple star must be at least several dozen AU apart. There are a few thousand of these star systems scattered across the sky, some making exquisitely beautiful sights in binoculars and small telescopes.

Sometimes, two stars of a binary are

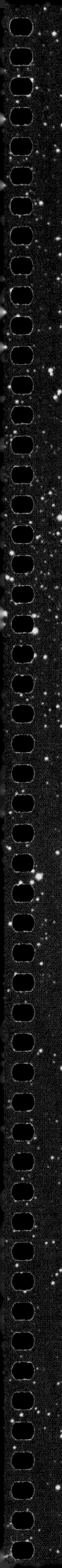

The Whirlpool Galaxy, M51, and its companion NGC 5195, **above**, *are seen as two smudges in small telescopes. The Whirlpool's spiral arms become visible in 10-inch and larger telescopes.* **Right**, *Zeta Orionis Nebula;* **previous page**, *Rosette Nebula.*

exactly the same brightness, while others differ slightly, and still others have wide variations in luminosity that contrast the two suns. However, the prizewinning doubles are the few with stars of completely different colours, indicating a wide difference in surface temperature. I enjoy looking at binary stars, knowing that I am gazing at two suns, each presumably capable of having planets. Earth could have been born into such a system. How different our sky would appear if a small red sun orbited in the place of Neptune.

The apparent distance between the components of double stars is measured in fractions of a degree. Remember that the moon is about half a degree in diameter. When viewed through a telescope or binoculars, the components of most double and multiple stars appear about as far apart as opposite sides of a small lunar crater. To define this distance, astronomers use minutes (') and seconds (") — one minute is one-sixtieth of a degree, and one second is one-sixtieth of a minute. The unaided eye can distinguish two equally bright stars 6' apart, about one-tenth of a degree. If the stars are closer than that, binoculars are needed. Their increased resolving power will separate doubles down to just under one minute.

A handful of the widest binary-star systems are within range of binoculars. A 60mm telescope vastly increases the number of multiple stars visible because it can separate two equally bright suns apparently only 2.0" apart. A 3-inch telescope under the best conditions will divide two stars 1.5" apart. Using such an instrument, Massachusetts amateur astronomer Glenn Chaple has observed more than 1,400 different double and multiple stars during the past two decades.

About 10,000 double-star systems are visible in a 6-inch telescope, which will resolve two stars of equal brightness 0.8" apart. This is an exceedingly small measure, comparable to the apparent width of a dime three miles away. Larger telescopes will resolve proportionally closer pairs. However, these are theoretical figures assuming excellent optics *and* perfect atmospheric conditions. And, most important, the limits also assume that an experienced observer is at the eyepiece.

Beginners should not expect to come

close to the telescope's theoretical limit, and indeed, that is not the point of double-star observing. Stars that are easily distinguished make prettier sights than the ones which are so close together that the images can barely be separated. With these factors in mind, I have selected the best double and multiple stars for typical amateur equipment — both binoculars and telescopes. These stars are specifically identified on the charts in this chapter.

Not all stars that appear as doubles are together in space. A chance alignment of stars is called an optical double. The technical designation for a pair of gravitationally associated stars is binary system, but these objects are usually simply called double stars.

The brightness of individual stars in binary systems is given to an accuracy of one-tenth of a magnitude. Initially, it may seem difficult enough to tell a second-magnitude star from a third-magnitude star, but with a little practice, it is possible to distinguish stars differing by only a few tenths of a magnitude, especially when they are fairly close to one another. Many of the naked-eye stars on the charts in this chapter have their magnitudes indicated to the nearest tenth. After a few nights' practice, it should not be difficult to see the difference between, for example, a star of magnitude 3.3 and one of 2.9. In addition to their magnitudes, the separation of the stars in a multiple-star system is tabulated in minutes or seconds. The brightest star is known as A, the second brightest B, the third brightest C, and so on.

VARIABLE STARS

Although our sun is in the minority, being a single star, its light output remains stable over long periods of time. For thousands of years, the sun has not varied by much more than two percent from its present brightness, and most other stars are similar, stable thermonuclear furnaces with uniform energy outputs. However, a few stars are passing through critical stages in their evolution, where their thermonuclear generators are shifting gears from one type of fuel to another. What happens at this stage varies from star to star, depending mostly on its mass, but some stars undergoing this transition can oscillate in brightness by a factor of 15,000 in a span of just one year. Astronomers are keenly interested in these variable stars. Understanding them may reveal considerable information about stellar evolution and about the onset of death among stars, since many variable stars appear to be nearing the end of their lives.

About a dozen variable stars are visible to the unaided eye. Nearly 100 can be identified with binoculars, and a backyard telescope brings thousands into view. David Levy of Tucson, Arizona, one of the world's most experienced variable-star observers, has made more than 25,000 individual estimates of the brightness of these fluctuating suns. He is one of several hundred amateur astronomers who observe variable stars and report their brightness estimates to the American Association of Variable Star Observers (AAVSO) in Cambridge, Massachusetts. The AAVSO, in turn, tabulates the observations and forwards the results to variable-star specialists at major observatories, who have neither the time nor the staff to keep a nightly tab on the sky's population of variables.

Apart from its opportunities for assisting in scientific research, variable-star observing is one of the best ways to increase the eye's sensitivity for distinguishing faint objects and for detecting brightness differences. Furthermore, the process of hunting for the stars significantly expands the observer's knowledge of the sky. I began observing variables in my second year of telescopic exploration. I made more than 1,000 brightness estimates in the next two years, but for me, the experience was most valuable as an apprenticeship in learning celestial geography.

A few of the brightest variable stars are labelled on the charts in this chapter. Stars of suitable comparison magnitudes located nearby are identified to one-tenth of a magnitude. To practise calculating a variable star's brightness, pick a star slightly brighter or fainter, then make an estimate of the variable's magnitude. Predictions of the maximum brightness of a few prominent variables are given in the *Observer's Handbook* (see Chapter 12).

There are four main classes of variable stars, all of which have examples indicated on the main charts. *Cepheid* variables, named after the prototype, Delta (δ) Cephei, are highly regular pulsating stars. Their period of variability and their range in brightness are so precise that they are used as "standard candles" to determine the distances to other galaxies.

Eclipsing variables are binary stars in which neither star varies, but their orbits alternately place one star in front of the other, thus producing an eclipse and a decrease in light. The star Algol in Perseus is the best-known member of this class.

Long-period variables are red giants, similar to Betelgeuse and Antares but at a different stage in their evolution. Some of them vary by more than 10 magnitudes in less than a year, and it is not unusual for these stars to change by one or two magnitudes in a few weeks. These are the favourites of backyard variable-star observers because of the wide magnitude range and because the cycles do not exactly repeat each other, introducing an element of anticipation into a night's observing.

The *irregulars* are the final type of easily observed variable star and include a variety of oddball stars, some that are normally bright and then become dimmer and others that are normally faint and occasionally brighten. Another group of irregular variables ponderously oscillates over months or years, brightening or fading by a few tenths of a magnitude. Betelgeuse is in this category.

The most dramatic variable stars are *novas*, which unpredictably blast off their outer layers, the explosion causing the star's brightness to shoot up by 12 or 15 magnitudes. Normally, these are totally obscure, very faint stars. The brightness surge occurs over a period of hours or days, so that in effect, a "new" star appears. The nova of 1972 in the constellation Cygnus was, at its brightest, almost equal to Deneb. Two nights later, it had dropped to third magnitude and, in a few weeks, was visible only with binoculars.

Because of the sudden appearance of these stars, the backyard astronomer who is familiar with the sky can be the first to spot one. Since novas are relatively rare (only one or two a decade reach third magnitude), a nova discovery is a major astronomical event.

Novas occur in close binary-star systems in which one of the stars is a dense white dwarf whose intense gravity is vacuuming up material from its companion. Eventually, the dwarf becomes overloaded, heats up and blasts off its captured star-stuff, producing a brilliance that lights up a sector of the galaxy for several weeks. The star returns to normal after a few months, and decades or centuries later, the cycle repeats itself. Only a few novas have cycles shorter than a human lifetime, and these are diligently watched by seasoned variable-star observers.

The rarest class of variable is the *supernova*, the sudden, explosive death of a massive star. A supernova occurs when a star's central thermonuclear furnace runs short of fuel and shuts down. The star collapses, but the heat generated by the collapse produces a ferocious fireball billions of times brighter than the sun. In a few hours, an apparently normal star can become almost as bright as a galaxy! The supernova slowly dims during the next year or two.

Superluminous stars like Rigel are supernova candidates. If Rigel became a supernova, it would be hundreds of times brighter than the full moon for weeks. But supernovas are so rare, the chances of one occurring as close as Rigel are exceedingly slim. The last known supernova in our galaxy was observed by astronomer Johann Kepler in 1604. It was as bright as Jupiter. In February 1987, a supernova erupted in the Large Magellanic Cloud, the Milky Way's small neighbour galaxy. Supernova 1987A, as it was called, reached third magnitude, but its location in the far southern sky made it invisible from anywhere north of Central America.

STAR CLUSTERS

In Chapter 4, we identified the Hyades and the Pleiades, in the constellation Taurus, two clusters of stars several hundred light-years away that are visible to the unaided eye. Hundreds of similar clusters are scattered around our galaxy. They range from modest collections of a few dozen suns to swarms of thousands of stars kept in a huddle by their mutual gravity. These aggregations are known as open or galactic star clusters, but more often, they are simply called star clusters. They offer one of the truly rewarding categories for deep-sky hunters, since each has a unique appearance. Photographs of a few star clusters are scattered throughout this book, and descriptions of several dozen more are included on the charts that follow.

Many star clusters are seen well in binoculars and usually look best through the telescope's lowest-power eyepiece. Because they cover a larger area of sky than most telescopes can take in, clusters such as M7, the Pleiades and the Beehive Cluster are seen better in binoculars.

The Pleiades

Four hundred light-years away is the most prominent star cluster in the sky — the Pleiades. This group is often mistaken for the Little Dipper because of the arrangement of its six brightest stars. The

Visible to the unaided eye as a pale, oval smudge in the Milky Way between the constellations Cassiopeia and Perseus, the Double Cluster is revealed in binoculars and telescopes as twin jewel boxes of sparkling stars.

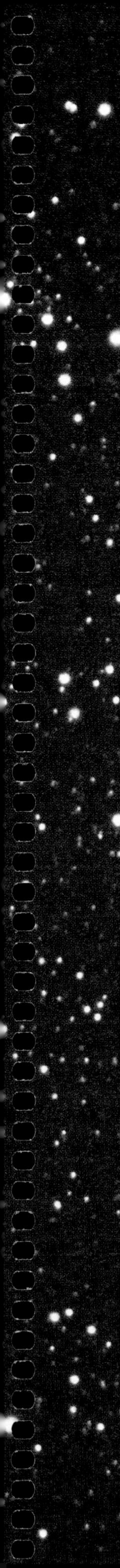

Pleiades are sometimes called the seven sisters, a reference to the seven stars visible to people with slightly better-than-average eyesight. I have a tough time seeing more than six stars with the unaided eye, even under excellent conditions, although some of my astronomy students report seeing as many as 11. Several dozen Pleiades stars can be detected in binoculars and a hundred or more by telescope.

The approximately 400 stars in this cluster were all born about 20 million years ago from a vast cloud of gas and dust, the remains of which are seen as wispy filaments in photographs (but they are faint and difficult to detect visually). Alcyone, the brightest of the Pleiades stars, is 500 times brighter than our sun. It shines with intense blue-white light, as do all the youthful stars in this cluster.

The Pleiades stars are so young — less than one percent the age of the sun — that the region can be regarded as a celestial nursery. The cluster is roughly 50 light-years in diameter. Near its centre, the stars average two light-years apart, about 50 times the stellar density in our vicinity.

In his classic *Celestial Handbook*, astronomer and historian Robert Burnham reports that in American Indian legend, the Pleiades are connected with Devils Tower, the curious, wonderfully impressive rock formation that rises like a colossal petrified tree stump 865 feet above the plains of northeastern Wyoming.

According to Kiowa lore, Devils Tower was raised up by the Great Spirit to protect seven Indian maidens who were being pursued by giant bears. The maidens were afterward placed in the sky as the Pleiades cluster, and the marks of the bears' claws may still be seen in the vertical striations on the sides of the Tower.

Certainly, the Pleiades star cluster has prompted inquiring gazes for many centuries. Chinese records include references to this celestial group as far back as 2357 B.C. However, if we could somehow be transported back in time to the age of the dinosaurs, the sky would not be adorned by the Pleiades. They would not be born for another 40 million years.

The Big Dipper Cluster

Studies of the motions of the seven Big Dipper stars over several decades have revealed that all but two of them are part of a star cluster. The two nonmembers are Dubhe and Alkaid, at the configuration's opposite extremities. The other five are all related. Astronomers estimate that they were born in the same region of space about 200 million years ago.

This group, plus about 30 other stars scattered across the sky, makes up the nearest star cluster. The reason it does not look like a star cluster is that we are inside it, although the sun is not a member. The Big Dipper cluster, or the Ursa Major cluster, as it is properly called, is slowly overtaking the sun, much like a group of joggers running in a clump gradually overtaking and passing a slightly slower runner. Millions of years in the future, the cluster will be an inconspicuous patch in the direction of the star Deneb.

Our descendants will witness the gradual breakup of the Dipper. The five cluster stars are moving in formation in the direction of the handle, while Dubhe and Alkaid are drifting the opposite way. This will bend the handle down at the end and open the bowl, but it will take many thousands of years. Our great-great-grandchildren will still see the same Dipper that we know today.

All the Dipper stars are far more luminous than the sun. Even the weakest, Megrez, is 20 times as luminous. If we were on a planet orbiting Megrez, the Earth's sun, 65 light-years away, would be an insignificant sixth-magnitude star. The beacon of the night would be Megrez's neighbour Alioth, only 7 light-years away, shining as brightly as Venus does in our sky. The other members of the Dipper cluster — Mizar, Phecda and Merak — would be celestial beacons ranging between Sirius and Vega in brightness.

But as star clusters go, the Dipper cluster is a loose aggregation. If the stars were once bound by mutual gravitation into a compact group like the Pleiades, the bonds have since loosened. Today, they are just floating in space alongside each other. In a few hundred million years, they will be completely dispersed, just as the cluster from which our sun formed lost its identity long ago.

NEBULAS

Drifting among the stars in the galaxy's spiral arms are thousands of vast clouds of gas and dust called nebulas. Most of them are dark and invisible, sometimes producing the dark rifts and patches that give the Milky Way its ragged, segmented appearance. These clouds are the largest objects in the galaxy, sometimes hundreds of light-years across — massive celestial smog banks that block the light of millions of stars. Generally, their presence is revealed only by a paucity of stars along certain sectors of the Milky Way. Occasionally, though, they produce something spectacular when stars illuminate the normally dark veils of gas and dust.

Nebulas are the galaxy's maternity wards, where new stars are born. The process begins inside the nebula, where gas and dust collect in knots, then disperse, only to collect elsewhere. Sometimes, the cloud's equilibrium experiences a major disturbance (perhaps due to a merger of two or more clouds), inducing pockets of cloud to collapse into denser clumps. Once a rapid infall of matter begins, a gravitational chain reaction is triggered. Atoms and molecules bang into each other and into the cloud's dust grains. Energized by such collisions, the dust radiates heat. But the opaque cloud traps the radiation, heating it further. Meanwhile, matter is piling up at the gravitational core of the pocket, and the temperature soars. The core mass attracts more matter, and the cycle escalates. In tens of thousands of years, a brief span of time by astronomical standards, temperatures climb from minus 250 degrees C to 15 million degrees, the ignition point for fusion reactions. A star is born.

The details of this process are still unclear, because the infalling envelope of gas and dust seals the stellar birth from view.

The infant star — called a protostar — hides in the womb of its mother cloud. Eventually, like afterbirth, the star rids itself of its cloak by the sheer force of its radiation. A section of the cloud is blown off, the youthful star exposed. The star will likely not be alone. The process that produces one usually acts on a substantial zone of the giant cloud, and a star cluster emerges as if bursting from a cocoon. This is exactly what is seen in several sectors of the Milky Way. The best-known and brightest of these star-birth regions is the Orion Nebula, just below Orion's belt.

The Orion Nebula

At the core of the Orion Nebula is a cluster of stars born about 50,000 years ago, according to current estimates. These are energetic suns, more massive and more luminous than our own. They have swept back the dust veils, illuminating an awesome 20-light-year-wide bowl-shaped cavern — the visible zone that astronomers call the Orion Nebula — on the edge of a titanic dark cloud that fills most of the constellation Orion.

More of the dark cloud will become exposed in the millennia ahead. Studies with infrared telescopes confirm that new stars are forming in the thick gas and dust immediately behind the Orion Nebula. As the new stars evolve and emit more heat and light, they will evaporate the dust and ionize the gas, causing it to glow — just as the stars at the core of the Orion Nebula have done. Star birth is like a spreading infection; once it gets started, the process eats away at the giant interstellar cloud.

The Orion Nebula is the nearest bright nebula, 1,600 light-years away, and the only one plainly visible to the unaided eye (from north of +20 degrees latitude). It looks like an out-of-focus star, a fourth-magnitude puff of cosmic cotton. Binoculars, however, show something more

The Lagoon Nebula (larger object) and the Trifid Nebula are a pair of star factories in the Sagittarius sector of the Milky Way. Both are visible in binoculars. Telescopes show the major details in this time-exposure photograph.

than a hint of the hazy patch of light: The soft glow offers a captivating contrast to the three second-magnitude stars in Orion's belt and the third- and fourth-magnitude stars closer to the nebula.

Binoculars also reveal a fifth-magnitude star near the nebula's centre. This is Theta 1 Orionis, one of the most intriguing star systems in the sky. Appearances are not deceiving here — Theta 1 *is* actually at the heart of the Orion Nebula and is the main source of its illumination. Even a 2-inch telescope discloses the unusual nature of this star: it is not a single star but four arranged in a trapezoid. Called the Trapezium, the four range in brightness from fifth to seventh magnitude and are far enough apart (the smallest separation is nine seconds of arc) to make an exquisite scene — four blue jewels embedded in a delicate celestial cloud.

Spectroscopic examination has revealed that all of the Trapezium stars are doubles, and other studies suggest that many of the surrounding stars are gravitationally associated with the bright four. So the Trapezium stars can be considered the brightest members of a small star cluster.

Photographs of the Orion Nebula capture the vast extent of the wisps and veils that cover an area almost a degree across. Yet an initial look at the nebula through a telescope may cause the observer to wonder whether this really is the same object that the photographs show. Why is it so faint? And where is the colour?

When you use the lowest-power eyepiece, the nebula — at first — looks like a small, moderately bright patch around the Trapezium. But if there is no moon and the sky is free from haze or artificial light, a glorious sight will unfold. As the eye becomes accustomed to the low light levels, subtle rifts of the nebula appear, just at the threshold of vision. The larger the telescope, the more that can be seen. But even a small refractor will give delightful views of the distant star factory.

The human eye, however, has limitations. Its colour sensitivity at low light levels is practically nil. To me, the nebula has never appeared anything but pale grey-green, even in a 16-inch telescope.

Dark-adapted eyes can detect curious ripples and loops in the nebula's brighter sections. Try higher powers on these regions; this is one of the few nebulas that do not become washed out at high power. In larger telescopes, the nebula's core is a magnificent sight.

Experienced deep-sky observers use a technique known as *averted vision* when observing the Orion Nebula or any object of low surface brightness. The concept of averted vision is to concentrate on the celestial object within the telescope's field of view without looking directly at it. Often, details invisible with direct vision suddenly become distinct. It works because of the extra sensitivity of the outer areas of the visual receptors — away from the central axis of vision. This sensitive peripheral vision, developed millennia ago by our ancestors (who must have needed it for self-preservation), should always be employed when observing objects at the threshold of vision — either stars or nebulas — with or without a telescope.

Another standard technique for examining nebulas and other faint, extended objects is the use of low power. Low power not only produces the widest field of view but also increases the effective surface brightness of the object, compared with higher-power viewing.

The Orion Nebula's cloudlike appearance in photographs is somewhat deceptive. One can almost imagine winging through it in a spacecraft, moving among its wisps and swirls like an airliner passing through cumulus clouds. The actual situation is very different: a hypothetical spaceship rocketing through the nebula would encounter only slightly more particles than those recorded in interstellar space. The average density of the nebula is one-millionth the density of a good laboratory vacuum. The density increases near the sites where new stars are just beginning their lives or where nascent stars are being nurtured. Yet the nebula is so vast, more than 2,000 cubic light-years, that it contains enough material to form hundreds of new suns.

The gigantic dark cloud from which the Orion Nebula and its new stars emerged is illuminated in another sector of the constellation, producing the spectacular Horsehead Nebula. The dark mass that forms the horse-head shape is silhouetted against the bright nebula behind it, which in turn is illuminated by a dazzling star. We get just a peek over the edge of the dark mass at what is likely another nebula comparable to Orion. Such circumstances are the rule, rather than the exception. Hundreds of bright blisters like Orion are undoubtedly obscured from our view by rifts of dark interstellar smog. Less than one-trillionth of a star's light output will penetrate from one side of a typical giant dark cloud to the other. In the case of the Horsehead Nebula, there is more dark than bright, making it completely invisible to the unaided eye and only dimly seen in the best backyard telescopes.

GLOBULAR CLUSTERS

More distant than the galaxy's open star clusters and nebulas are globular star clusters, which are, in effect, tiny satellites of the Milky Way Galaxy. At least 140 globular clusters surround the galaxy, about one-third of which are visible in backyard telescopes. Omega Centauri, at fourth magnitude, is the brightest of these and is easily visible to the unaided eye. Unfortunately, because of its location in the southern sky, it is visible only from the extreme southern United States or farther south. The next brightest are the globulars M13 and M22, in Hercules and Sagittarius, both easy binocular targets.

These great spherical swarms of stars range from 50 to 200 light-years in diameter and have populations of up to two million suns. Backyard telescopes less than 4 inches in aperture will show them as concentrated balls of light gradually fading off at the edges. Four-inch and larger instruments will resolve some of the brightest stars. Averted vision improves the view substantially. Globular clusters benefit from increases in telescope aperture more than any other class of celestial object. In large telescopes, they are stunningly dramatic sights.

GALAXIES

Everything described so far is part of the Milky Way Galaxy. The final category of deep-sky objects is other galaxies — distant islands of billions of stars that float in the cosmic emptiness out to the greatest distances that telescopes can penetrate.

There are two classes of galaxies available to backyard astronomers: spiral and elliptical. Spiral galaxies are similar to the Milky Way. Some have their twirling arms tightly wound, producing almost uniform discs, while others have loose, ragged configurations. Spiral galaxies come in a limited range of sizes, from systems of a few tens of billions of stars to giants with several trillion. Many spiral galaxies are approximately the size of the Milky Way.

Elliptical galaxies, basically featureless spherical systems of stars, have a far greater range of masses, from dwarf ellipticals with a few million stars to the titanic supergiant elliptical galaxies containing up to 100 trillion stars. These are the largest objects that backyard telescopes can detect. Some can be seen at distances of 100 million light-years or more.

As with nebulas, the details that long-exposure photographs reveal in galaxies can never be seen with the eye. However, the shape of a spiral galaxy is often evident. In general, a bright nucleus that appears like a star embedded in a small patch of mist is surrounded by a fainter envelope of light. Occasionally, there is a hint of spiral arms, but more often, the spiral galaxy will simply be a general haze around the brighter nucleus. Edge-on spirals look like delicate slivers of light. The two nearest galaxies, the Large Magellanic Cloud and the Small Magellanic Cloud, are deep in the southern sky and

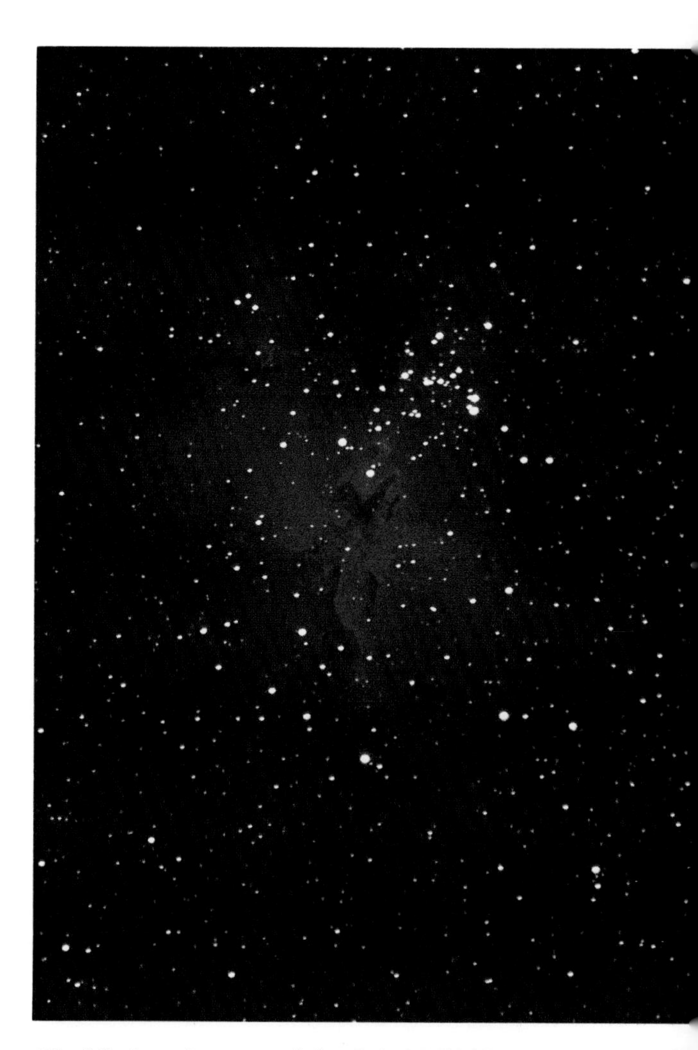

Huddled at the core of the Orion Nebula are four stars known as the Trapezium, **left**, *less than a light-year apart but plainly visible in small telescopes.* **Above:** *The Eagle Nebula.*

are not visible from anywhere in Canada or the United States.

The nearest galaxy similar to our own is the Andromeda Galaxy, seen as a hazy fourth-magnitude object near the star Nu (ν) Andromedae. At two million light-years, the Andromeda Galaxy is the most distant object that the unaided eye can detect. Binoculars reveal its oval outline, about three degrees long by less than one degree wide. Because the galaxy is tipped only 18 degrees to our line of sight, it has an oval shape. The galaxy's two small elliptical companions, M32 and NGC 205, are visible in 4-inch and larger telescopes. In general, though, the Andromeda Galaxy is a disappointment for backyard astronomers. It looks so much more mag-

Photographs usually show galaxies and other deep-sky objects far brighter and in greater detail than they appear in backyard telescopes. However, the photograph of the galaxy NGC 253 and the globular cluster NGC 288, **above**, *comes close to portraying the typical telescopic appearance of these two classes of objects. At low light levels, the eye is largely insensitive to colour. Galaxies appear as grey, oval smudges, and star clusters are swarms of silvery points of light. When observing galaxies, remember that individual stars seen near them are members of our own Milky Way Galaxy and are not associated with those remote star cities. Like raindrops on a windowpane, they speckle the distant scene. Conversely, open clusters,* **left**, *are inside our galaxy and therefore at about the same distance as many of the surrounding stars in the photograph. The clusters are, from left to right, M37, M36 and M38 in the constellation Auriga. All three are visible in binoculars and are fine telescopic targets.*

nificent in photographs. At least substantial structure is seen in the Orion Nebula, which also produces spectacular photographic portraits.

In many instances, galaxy observers will be content with just seeing these remote cities of stars, their pale, delicate forms stimulating the mind more than the eye. For example, consider M51, the Whirlpool Galaxy, which is located about a fifth of the way from Alkaid in the Big Dipper to Cor Caroli in the small constellation Canes Venatici. Although equal in size to the Milky Way Galaxy, M51 is 35 million light-years away and shows up as merely a faint smudge in small telescopes. Telescopes over 4 inches in aperture may show it as a double smudge.

The Whirlpool Galaxy is one of the few galaxies whose spiral structures can be glimpsed with typical backyard astronomy equipment. I have never seen it clearly with anything less than a 10-inch; but some amateur astronomers using 5-inch telescopes have picked out one or two spiral arms. The magnificence of this continent of stars, however, can be seen only in long-exposure photographs, which also reveal that the "double smudge" effect is due to a companion galaxy apparently dangling at the end of one of M51's spiral arms. The companion, known as NGC 5195, is another spiral galaxy behind M51 that was severely distorted a few hundred million years ago during a close encounter with M51. This galactic sideswipe wrecked the shape of NGC 5195 and distorted M51's symmetrical spiral.

Such near collisions are not uncommon in the realm of galaxies. What is unseen, however, even in photographs, is that many billions of stars must have been wrenched away from both of these galaxies and flung into the abyss of intergalactic space. If our galaxy suffered a similar experience and our sun were torn away from its parent galaxy, Earth would remain unaffected, continuing in solar orbit while the sun drifted forever in the intergalactic darkness. The sky would be almost totally black, punctuated here and there by hazy islands — the nearest of the universe's billions of galactic star cities.

TELESCOPE EXPERIENCE

It takes a long time to get used to looking through a telescope, especially when the target is a dim deep-sky object. Some of my astronomy students are unable to see anything when they step up for their first look. The biggest problem seems to be keeping the head steady while hunching over to look in the eyepiece. Observing with a telescope means training the body as well as the mind. It requires that the observer stand rigidly still but relaxed. The eye and the mind must slowly be trained to pick out details that are initially imperceptible. The telescope's lowest magnification should always be used at first. This makes it easier to locate, focus and see the celestial target clearly. (Trying to avoid the cool evening air by pointing a telescope through a window is useless. The window glass introduces severe distortion to astronomical views. Opening the window will not work either. The flow of air through the opening generates very poor seeing.)

The delicate spindle shape of an edge-on spiral galaxy or the filamentary tendrils of a distant cloud of dust and gas like the Orion Nebula are sights that are almost invisible to the novice. After several weeks, my students are usually able to see details completely undetected when they first looked through the telescope. But they are awed by the fact that I can aim the telescope at an apparently vacant part of the sky, twiddle a few knobs and present to them a star cluster or a nebula. This is really not a special talent. All veteran backyard stargazers can do it. But it is largely a self-taught skill.

Being able to locate objects in the sky is primarily an exercise in self-edification. Someone can point out the main constellations, but when it comes to seeking the fainter objects buried among the myriad stars, every backyard astronomer must serve an apprenticeship of self-taught sky knowledge. Once the geometrical relationships of the constellations begin to link in your mind, the celestial clockwork — the sky motions caused by the Earth's rotation on its axis and by its orbit around the sun — will soon become familiar. After a year or two, it all starts to fall into place, and the night sky becomes more than just a pretty tapestry of stars.

Eventually, most amateur astronomers come to know by memory the relative positions of the brightest 500 to 1,000 stars. That provides a celestial web from which the more interesting telescopic quarry can be tracked. This sky-familiarization process is where many beginners give up. They pack up their telescopes because they are unable to find the celestial sights. But they only shortchange themselves by limiting the depth of the hobby and underestimating their own abilities.

Others, expecting to see Technicolor vistas like the observatory photographs displayed in coffee-table astronomy books, are disappointed when confronted with the real objects observed through backyard telescopes. But it is seldom pointed out in those books that if one could look through the telescope used when taking the photographs, the objects would appear much less impressive. Photographic film accumulates light from a celestial source for minutes or hours, whereas the human eye generally forms a new image in the brain every one-fifth of a second. Nevertheless, a look at the subdued, but real, image of a remote galaxy or nebula has a chilling quality that must be experienced to be appreciated.

DIAL-A-GALAXY

Many telescopes are equipped with two numbered dials, called setting circles, on the rotational axes. These are for correlating the aiming of the telescope with the sky coordinate system, which is analogous to latitude and longitude on Earth. A celestial equator and pole, directly above their Earthly counterparts, are the keys to the grid system.

Navigating the celestial coordinate system seems like the logical way to track down sky objects, yet relatively few amateur astronomers use this method. Significantly, the more experienced the backyard astronomer, the less the setting circles are used. To use the setting circles, the observer looks up the coordinates (known as right ascension and declination) of the desired object, then adjusts the telescope to the correct readings on the dials and, finally, peers through the eyepiece to see whether the celestial target is in view.

There are two problems with this, one philosophical and one practical. The practical problem is that the telescope has to be precisely aligned to the celestial pole and the setting circles must be accurate. With practice, aligning the telescope does not take much time, but the setting circles on many amateur telescopes are not precise enough to get the desired object in the field of view every time.

Recently, however, the computer age has surmounted this shortcoming with digital read-out setting circles (about $500) that are more accurate than mechanical setting circles and easier to use. Certain models are even smart enough to compensate for errors in polar alignment, though the observer must initialize the setup by pointing the telescope at two stars from a list provided by the manufacturer.

Even more sophisticated are computer-aided telescopes introduced in the 1990s, with the computer built in. The computer tracks the stars to compensate for the Earth's rotation and can access a database of hundreds of celestial objects that can be called up on a control panel. Once an object is selected, the computer uses the two axis motors to slew the telescope to point at the object's position. Such an arrangement may sound ideal for the beginner who has yet to learn the sky. But this is where I have the philosophical problem I mentioned earlier.

The challenge of hunting down celestial quarry using the eye, finderscope and telescope is pure backyard astronomy. Circumventing the hunt is like wearing a bag over your head while under the stars, only to remove it to look in the eyepiece. If experience at tracking down faint galaxies and nebulas is bypassed, the night sky never becomes the comforting dome of

The Andromeda Galaxy, **above**, *the nearest galaxy similar to our Milky Way Galaxy, is seen with its companions M32 (top) and NGC 205. All three, along with one or two dust lanes, can be detected with small telescopes. Compare this photograph with the one on page 18.* **Right:** *The globular cluster M22 in Sagittarius.*

familiar pathways that the true backyard astronomer knows it to be. The smudge of a remote galaxy means more when *you* find it, instead of some computer.

I searched nearly an hour one summer for the globular cluster M3 when I began tracking down telescopic targets. Now, it takes only a few seconds. However, if I had always been doing it using setting circles or a computer-aided telescope, it would *still* take the same amount of time working through a basically mechanical ritual. By using the visual-sighting method, I have come to know the sky in a way that would not be possible otherwise.

Although I strongly urge novice telescope users not to shortchange themselves by taking the seemingly easy computer-guided route to the galaxies, digital setting circles and computers do have their place. Several of my colleagues, all experienced amateur astronomers, say these accessories have allowed them to pursue more specialized and productive observing programmes. But beginners could more wisely invest equivalent dollars in a better telescope or additional eyepieces.

KEEPING RECORDS

No formalities or standard formats are needed for recording a night under the stars. Any type of notebook will do, although I prefer the spiral-bound high school workbooks with blank pages for quick sketches facing lined pages for notes. The important point is to begin keeping notes from the first night that something is recognized. Record the date, time, place, instruments used (if any) and objects seen. If something was searched for but not found, note that too.

Observing conditions at your site influence what can and cannot be viewed on a given night. Hazy skies work against detection of fainter objects but do not significantly affect observation of lunar craters, bright planets or double stars. To grade the clarity of the night sky, note the faintest star seen with the naked eye near the Little Dipper (Ursa Minor). Stars to sixth magnitude are marked for this purpose on Chart 1 and Chart 20. This provides a standard basis for comparison from one night to the next. After you have logged a few hundred objects, the book becomes a personal record of ever deeper penetration of the cosmos.

USING YOUR NIGHT EYES

Although humans cannot match the superb night vision of owls, cats and other nocturnal creatures, our eyes are remarkably efficient at seeing detail in the dark. An elaborate system for seeing in the dark evolved millions of years ago, when our ancestors had to perceive potential danger after nightfall. The process is called dark adaptation.

In a darkened environment, the eye reacts almost immediately by increasing the size of the pupil, thus allowing more light into the eye. This is the same as opening the f-stop on a camera from f/8 to f/2. Then, during the next 15 minutes or so, a more complex process takes place. The supply of visual pigment in the photoreceptors of the retina steadily increases, and as long as no bright light enters the eye, the sensitivity to dim light levels surges. This pigment-sensitizing process can be likened to changing the film in a camera from ASA 25 to ASA 4000.

Complete dark adaptation has astonishing consequences for stargazers. On a black, moonless night, the Milky Way stands out as a swath of light. Whereas it is invisible to someone just stepping out of a normally illuminated house, the dark-adapted skywatcher sees a profusion of faint stars, producing the illusion that the sky is covered with them.

Fully dark-adapted eyes can see about 3,000 stars in a dark sky well away from city lights. That may not seem like many, but compared with the 200 or so seen from typical well-lit suburban communities, it is almost the proverbial difference between night and day.

USING THE DEEP-SKY CHARTS

The 20 charts on the following pages present in detail almost the entire sky visible from midnorthern latitudes. Beyond their function as a star atlas, the charts provide information about hundreds of naked-eye, binocular and telescopic objects. Each chart includes one or two key constellations in an area of the sky about 45 by 55 degrees. All stars to fifth magnitude are shown. The Bayer or Flamsteed designations of brighter stars are given, along with magnitudes to the 10th. Other information, such as star diameter, luminosity and distance, should be viewed as increasingly less accurate the more remote the object. The reliability of deep-sky data depends primarily on the distance determinations. Distance estimates are generally very accurate out to 200 light-years, fairly good to 400 light-years and subject to chains of educated guesses after that. Information is colour-coded:

Black = Star and constellation names; constellation outlines.

Red = Descriptive astrophysical information, such as distance, size, luminosity, classification.

Blue= Observing information, such as type or class of object, magnitude, double- and multiple-star data, general appearance, instrumentation needed.

When you are using a red-filtered flashlight outdoors, the red lettering becomes almost invisible. This information is not required during outdoor observing.

Abbreviations:

ly = Distance in light-years.

Lum. = Luminosity compared to our sun.

Dia. = If coded red, diameter of star in solar diameters; if coded blue, apparent diameter in seconds or minutes, as indicated.

Blue SG = Blue supergiant, the most luminous type of star. These are massive high-temperature suns that burn so furiously, they last only a few million years.

Red SG = Red supergiant, the largest type of star; believed to be a late-life transition stage of blue supergiants. Final stage may be a supernova explosion.

Blue G = Blue giant, a less massive, less luminous, longer-lived version of the blue supergiant.

Yellow G = Yellow giant, a former blue giant that is evolving toward a red-giant stage.

Red G = Red giant, a star, usually more massive than the sun, in the last stages of its life.

STAR BRIGHTNESS CENSUS

Magnitude	Number of Stars per Magnitude	Total
−1 (−1.5 to −0.4)	2	2
0 (−0.5 to 0.4)	6	8
1 (0.5 to 1.4)	13	21
2 (1.5 to 2.4)	72	93
3 (2.5 to 3.4)	190	283
4 (3.5 to 4.4)	617	900
5 (4.5 to 5.4)	1,956	2,856
6 (5.5 to 6.4)	5,586	8,442
7 (6.5 to 7.4)	15,257	23,699

DESIGNATIONS OF SKY OBJECTS

The all-sky charts in Chapter 4 provided the names of the constellations and a few of the brighter stars. The charts in this chapter show many more stars, as well as the locations of a variety of deep-sky objects. Since only a hundred or so of the brightest stars have been assigned individual names (Vega, Arcturus, et cetera), there are other naming systems to identify the less prominent stars.

The oldest systematic attempt at star naming was developed by German astronomer Johann Bayer at the beginning of the 17th century, just before the invention of the telescope. Bayer designated the stars in each constellation in order of brightness by using lowercase letters of the Greek alphabet: alpha (α) for the brightest, beta (β) for second brightest, gamma (γ) for third brightest, and so on. For some of the larger constellations, Bayer used the entire 24-letter Greek alphabet.

By the end of the 17th century, astronomers realized that they had to remedy the limitation of Bayer's system. British astronomer John Flamsteed suggested assigning a number to each star in a constellation, thus eliminating the confining Greek alphabet. He applied numbers from the western side of a constellation toward the east, including each star within the limit of naked-eye visibility. The largest constellations received more than 100 of Flamsteed's numbers. But the Flamsteed system did not supersede the Bayer designations, and today, Flamsteed numbers are used only on the stars not covered by the Greek letters. Stars too faint for the Flamsteed list are each identified by one of several more recent catalogue numbers generated at observatories specializing in star positions. For example: BD36°2516, from the German *Bonner Durchmusterung* star catalogue.

Deep-sky objects have their own identification system, usually derived from the catalogue of 18th-century astronomer Charles Messier or from the *New General Catalogue* compiled by English astronomer J.L.E. Dreyer and published about a century ago. The 109 objects in the *Messier Catalogue* are designated M1, M2, and so on. The several thousand *New General Catalogue* objects are prefixed by the letters NGC. Most of the Messier objects are in the *New General Catalogue*, many of them having a popular name as well. M1, for example, is known as NGC 1952 and also as the Crab Nebula. A few objects not in either of these major catalogues have different prefixes, such as IC or Col., identifying other lists.

The Greek Alphabet

α	Alpha	AL-fuh
β	Beta	BAY-tuh
γ	Gamma	GAM-uh
δ	Delta	DELL-tuh
ϵ	Epsilon	EPP-sill-on
ζ	Zeta	ZAY-tuh
η	Eta	AY-tuh
θ	Theta	THAY-tuh
ι	Iota	i-OH-tuh
κ	Kappa	CAP-uh
λ	Lambda	LAM-duh
μ	Mu	mew
ν	Nu	noo
ξ	Xi	zeye
ο	Omicron	OHM-ih-krawn
π	Pi	pie
ρ	Rho	row
σ	Sigma	SIG-muh
τ	Tau	taw
υ	Upsilon	UP-sih-lon
φ	Phi	fie
χ	Chi	kie
ψ	Psi	sigh
ω	Omega	oh-ME-guh

STAR DIAMETERS

Since stars are too remote for any telescope to measure their diameters directly, how do astronomers know that Sirius is 1.8 times the diameter of our sun and that Aldebaran is 45 times wider? The answer is *indirect* measuring techniques, which have yielded the diameters of about 100 stars.

The first method involves the precise monitoring of a star when it is blocked out by the moon. As the moon makes its monthly orbital trek around Earth, it passes in front of many stars, but only occasionally are the stars bright enough for the detection equipment to complete the experiment. Although the technique is little more complicated than observing the length of time it takes the star to disappear (which is almost instantaneous), it has revealed accurate diameters for several dozen stars.

However, the moon's path is limited to a specific sector of the sky, and astronomers anxious to determine the sizes of other stars developed two exceedingly painstaking techniques — called stellar interferometry and speckle interferometry — that ultimately required years of refinement to capture the diameters of a handful of stars. Both methods involve electronic analyses to cancel out the interference of the Earth's atmosphere and to take advantage of the physical properties of light propagation. To appreciate the difficult nature of these experiments, consider the fact that a typical star seen from Earth is about the same size as a walnut on the Empire State Building viewed from a distance of 400 miles, or as far away as Toronto.

The techniques of stellar interferometry, speckle interferometry and electronic monitoring of lunar occultations have produced the diameters of some of the stars mentioned in this book. The diameter of a star can also be inferred from stellar-evolution theory, based on the calibrations provided by the diameters of stars that have been measured. In this way, almost any star can now be tagged with an estimated diameter.

In tallying up the diameters, we find that among all the naked-eye stars, less than one percent are smaller than our sun. And yet a census of *all* stars shows that the average star is both smaller and dimmer than the sun. Thus the incongruity of the starry sky as seen from the backyard: It is the giants, the blazing beacons of the galaxy, that make up the familiar constellations. The average citizens of our starry vault make a negligible contribution to the night sky.

Seen from 40 million light-years away, the Milky Way Galaxy and its neighbour, the Andromeda Galaxy, would resemble M65 and M66, the brightest of this trio of galaxies in Leo.

Spring
Summer
Autumn
Winter
INDEX FOR ATLAS CHARTS
1
2
3
4
5
6
7
8
9
10
11
12
13
14
15
16
17
18
19
20

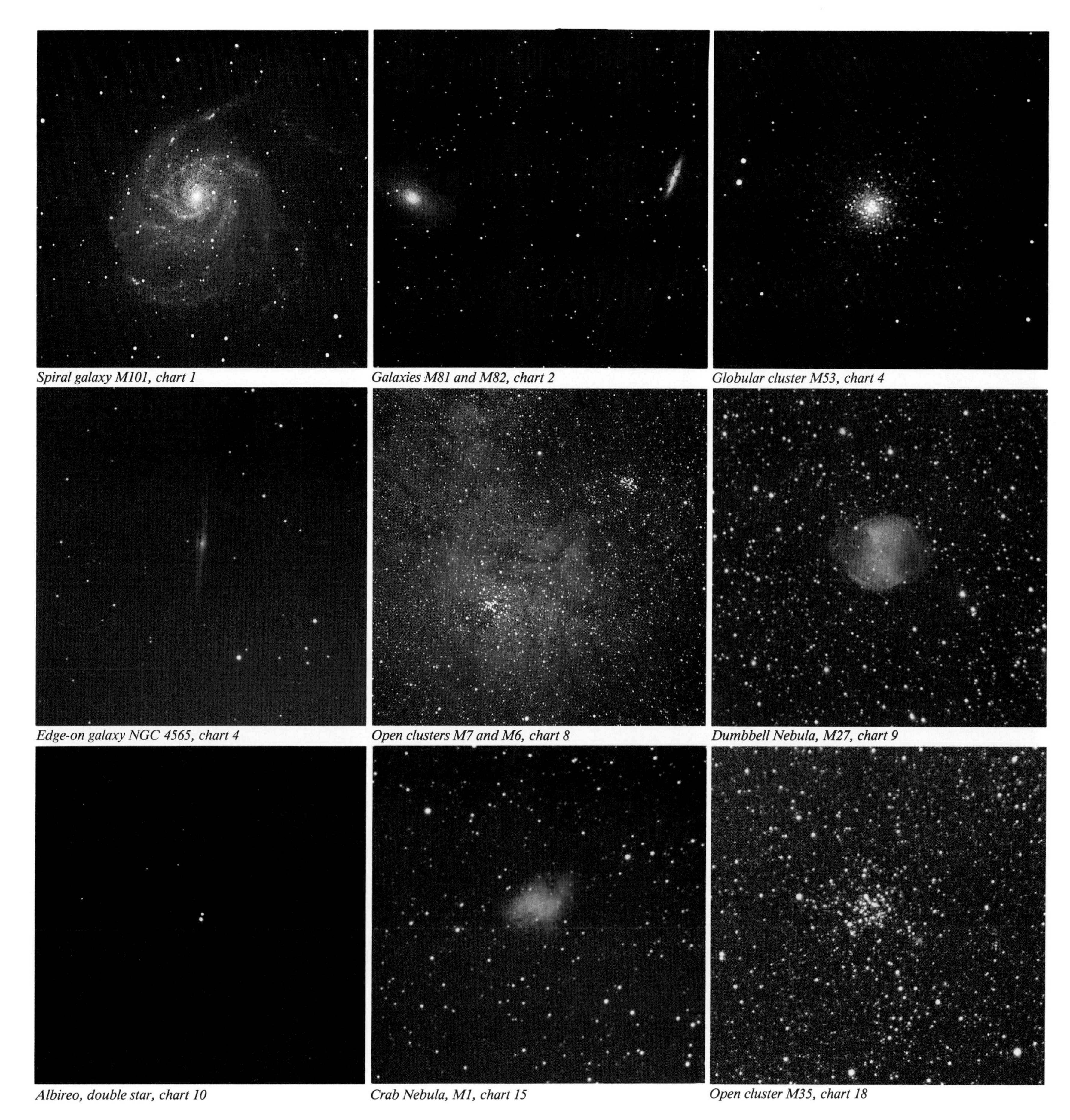

Spiral galaxy M101, chart 1

Galaxies M81 and M82, chart 2

Globular cluster M53, chart 4

Edge-on galaxy NGC 4565, chart 4

Open clusters M7 and M6, chart 8

Dumbbell Nebula, M27, chart 9

Albireo, double star, chart 10

Crab Nebula, M1, chart 15

Open cluster M35, chart 18

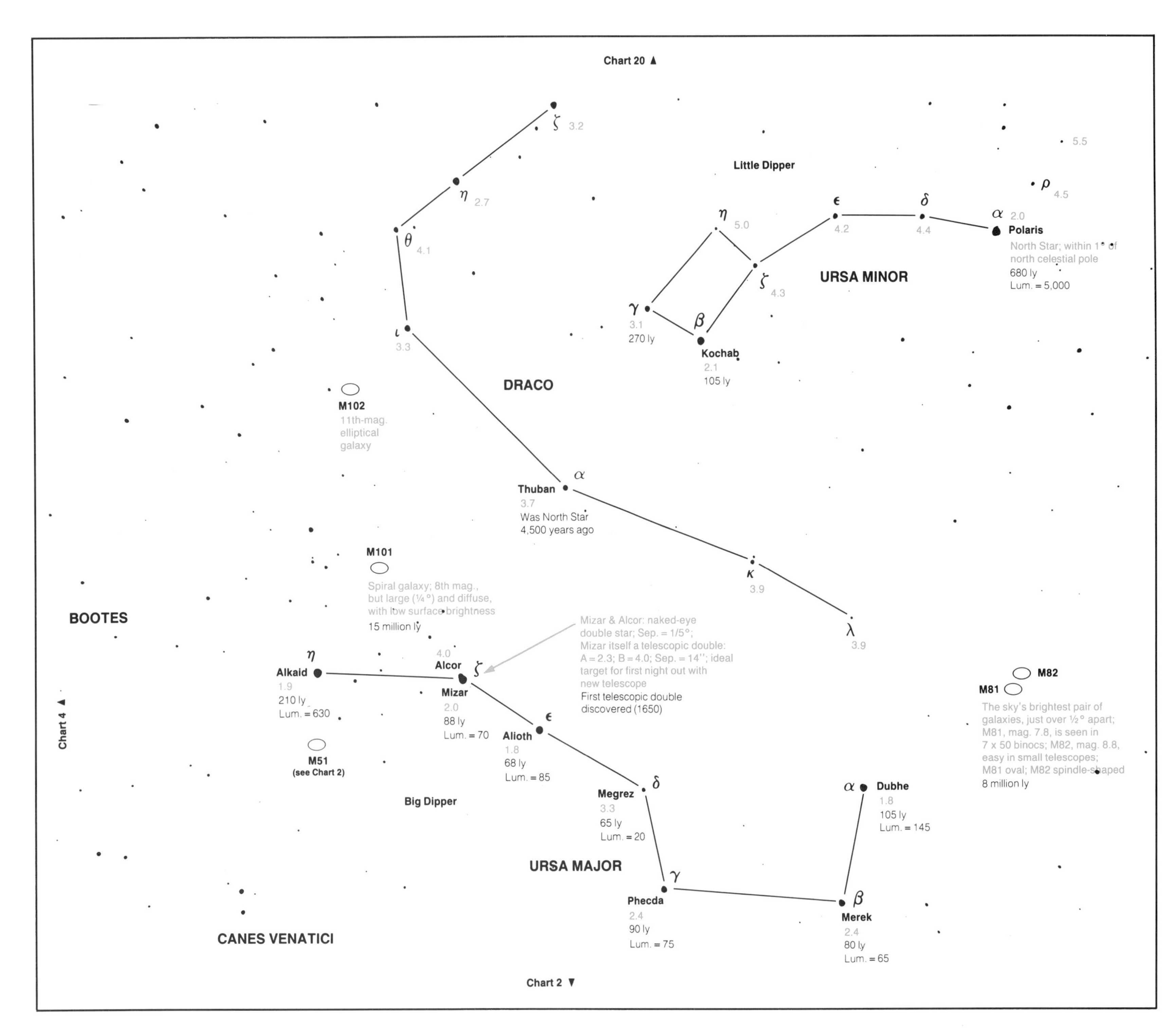

CHART I

Ursa Major
Ursa Minor
Draco

Visible all year in north; best orientation in late winter and spring.

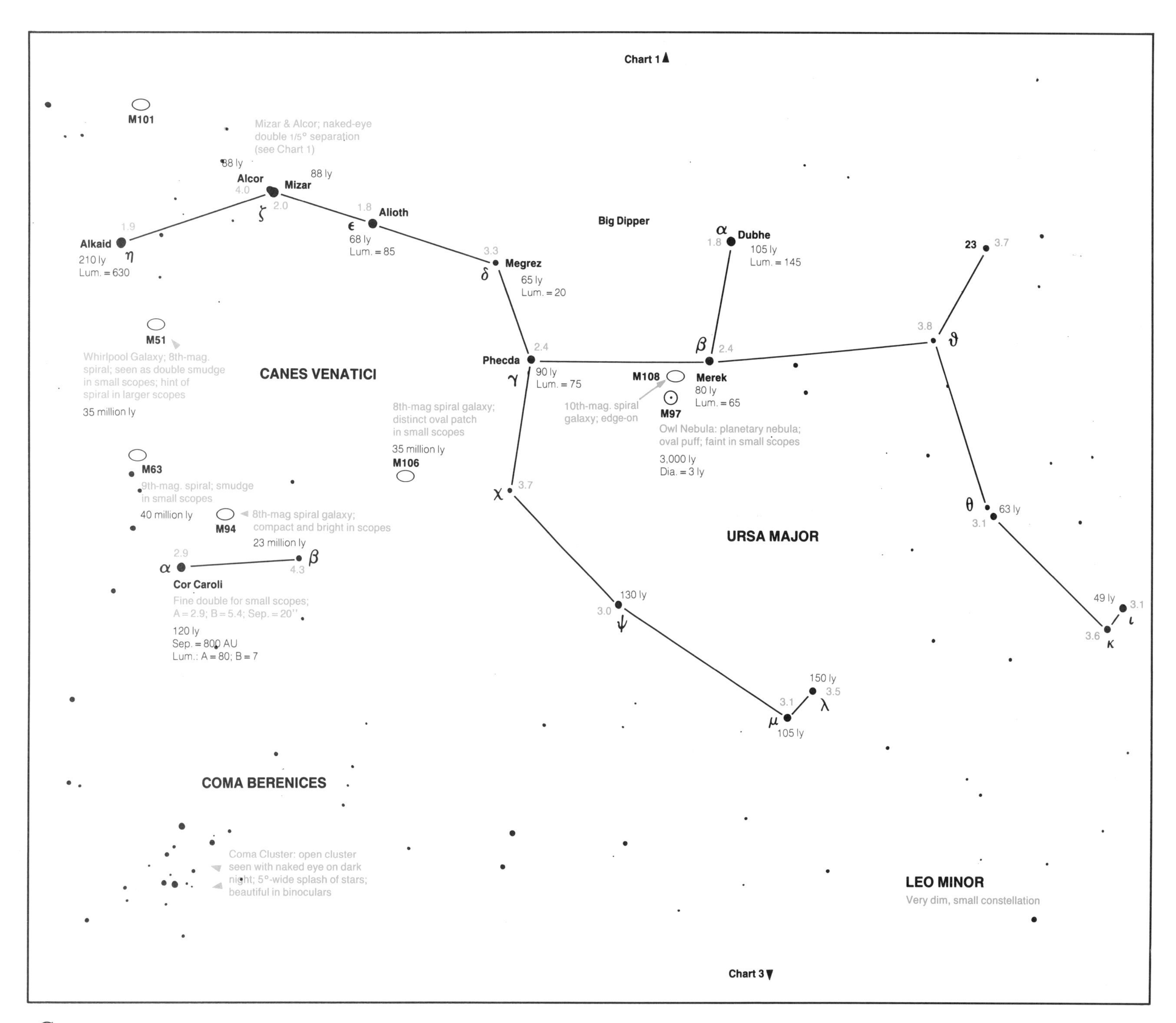

CHART 2

Ursa Major
Canes Venatici
Coma Berenices

Well positioned for viewing in late winter, spring and early summer.

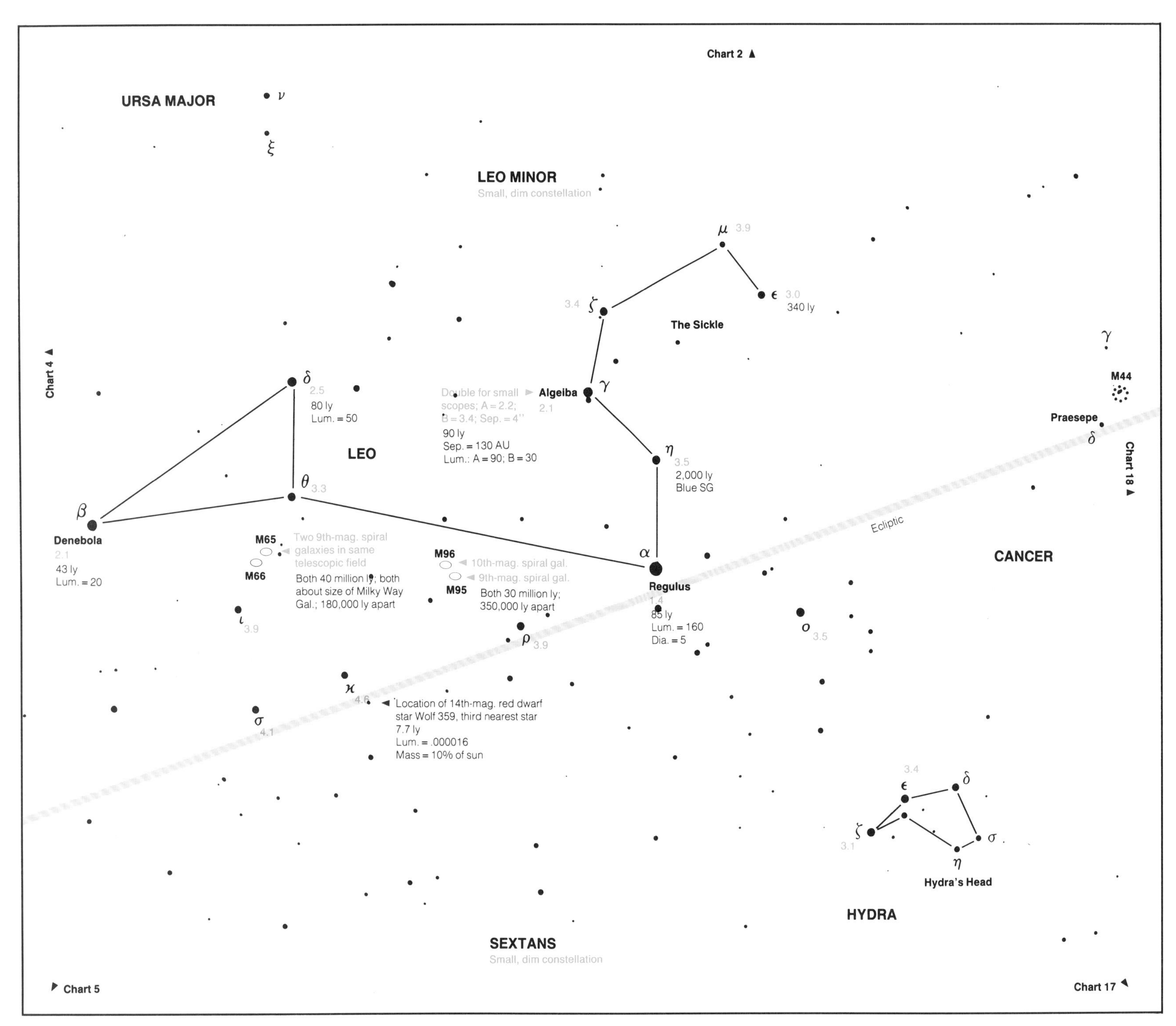

CHART 3

Leo
Cancer
Hydra

Prominently placed in spring and early-summer skies.

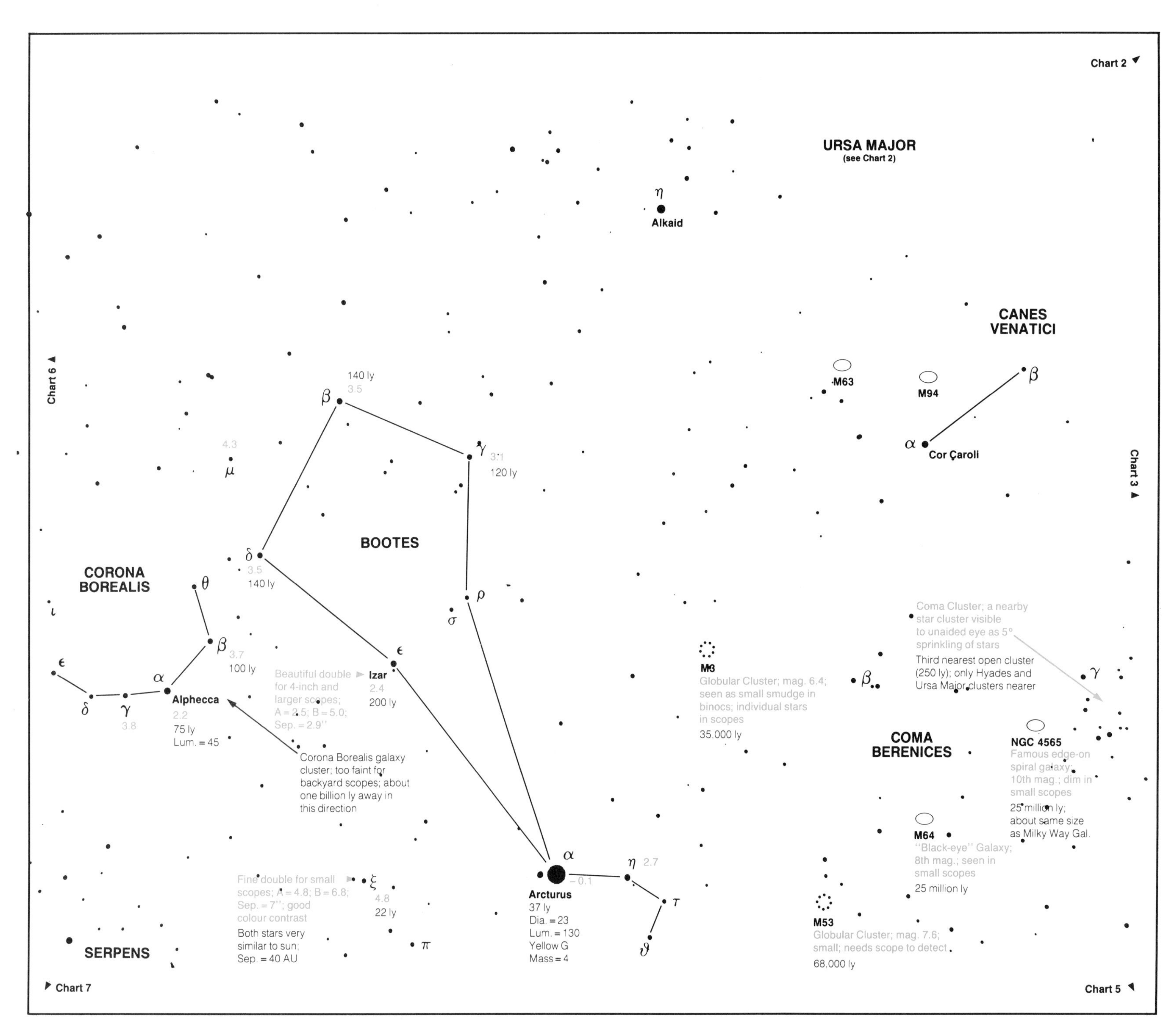

CHART 4 **Bootes**
Corona Borealis
Coma Berenices

Region high overhead in spring and early summer.

Chart 5

Virgo
Corvus
Hydra

Well placed in spring skies; bottom of chart close to southern horizon.

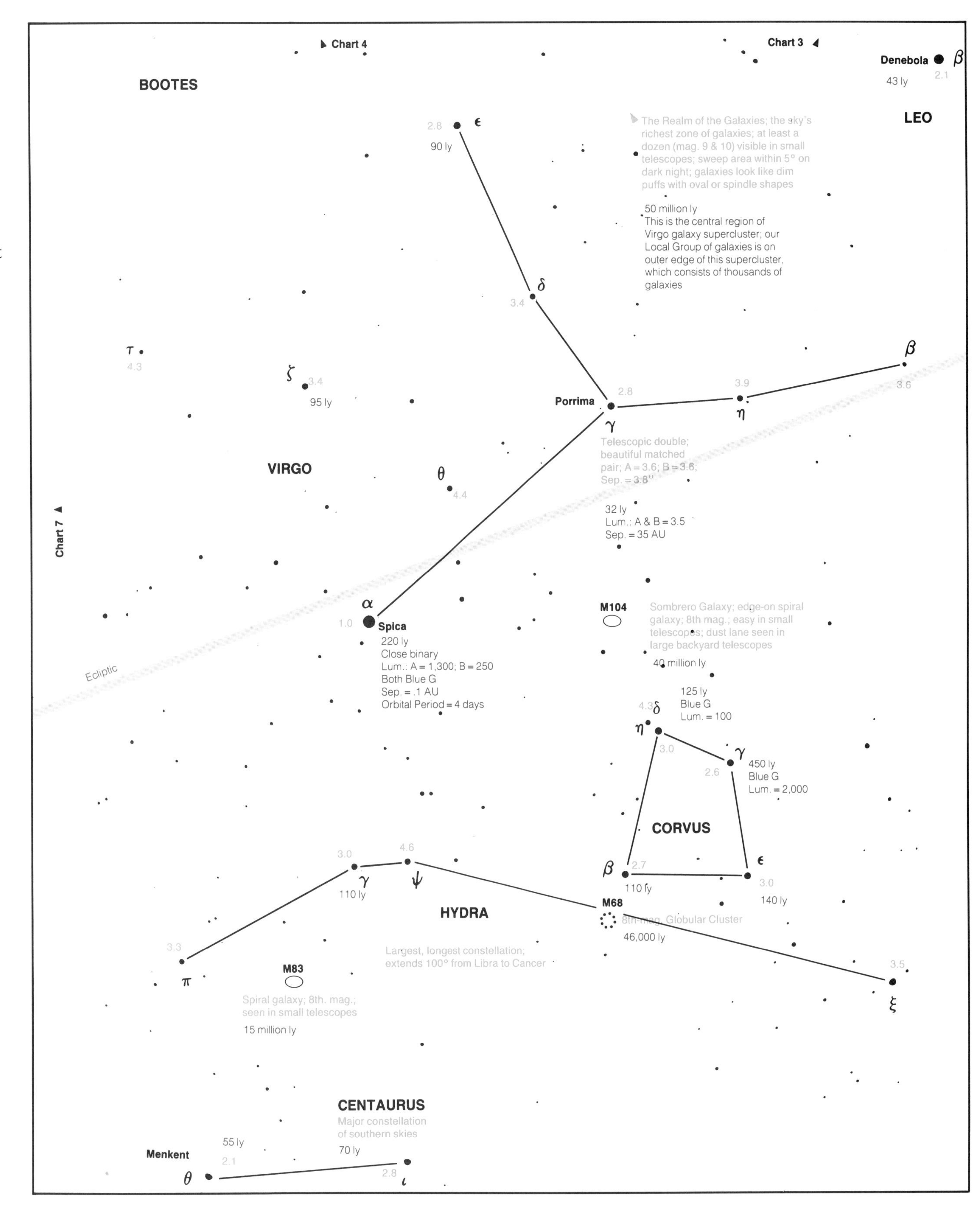

Chart 6

Hercules
Ophiuchus
Draco (head only)

Near overhead in late spring and throughout summer.

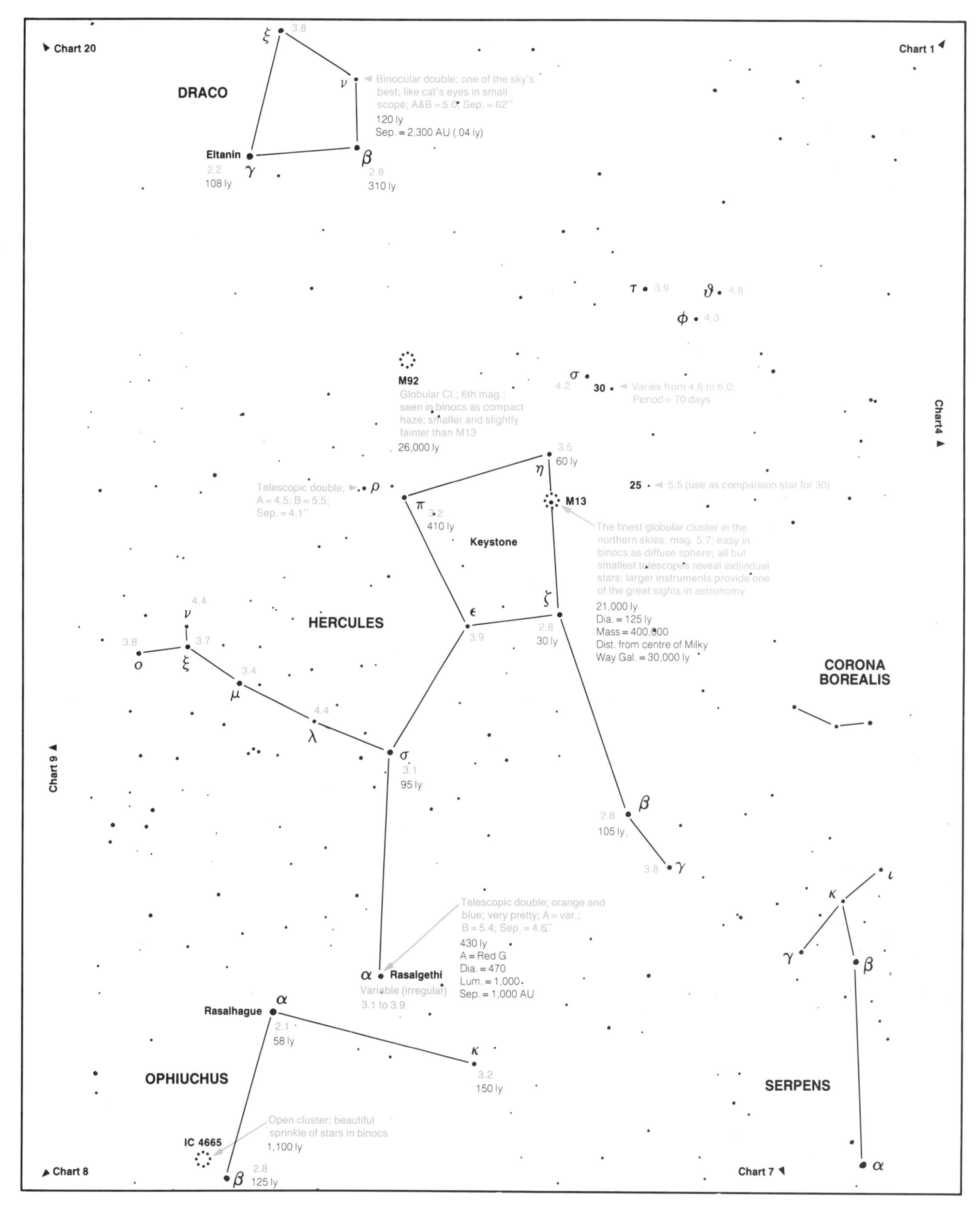

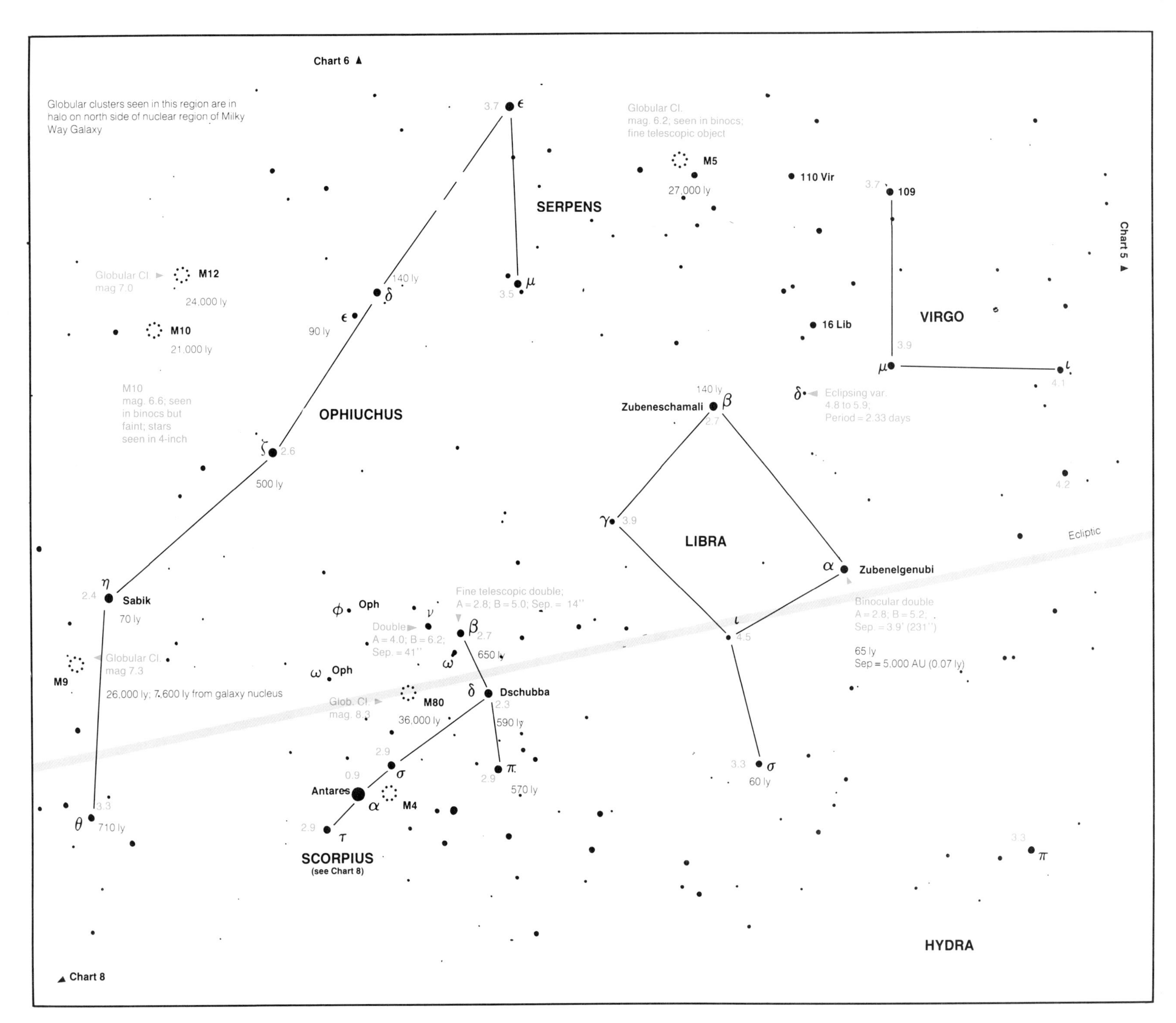

CHART 7

Ophiuchus
Libra
Scorpius
(northern part)

High in southern sky throughout the summer months.

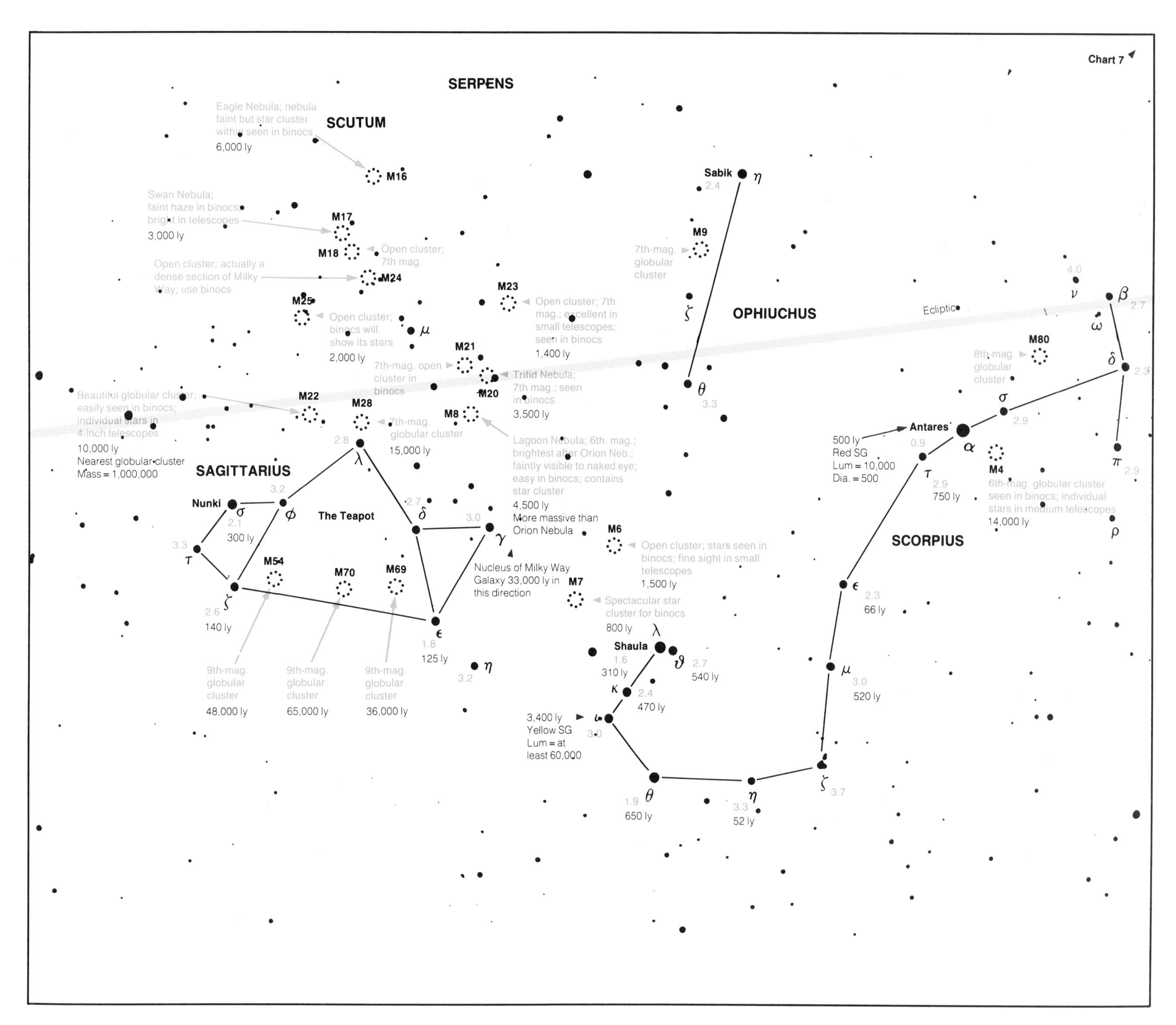

CHART 8

Scorpius
Sagittarius
Scutum

Richest zone of Milky Way carves through left side of this region; seen near south horizon mid- to late summer.

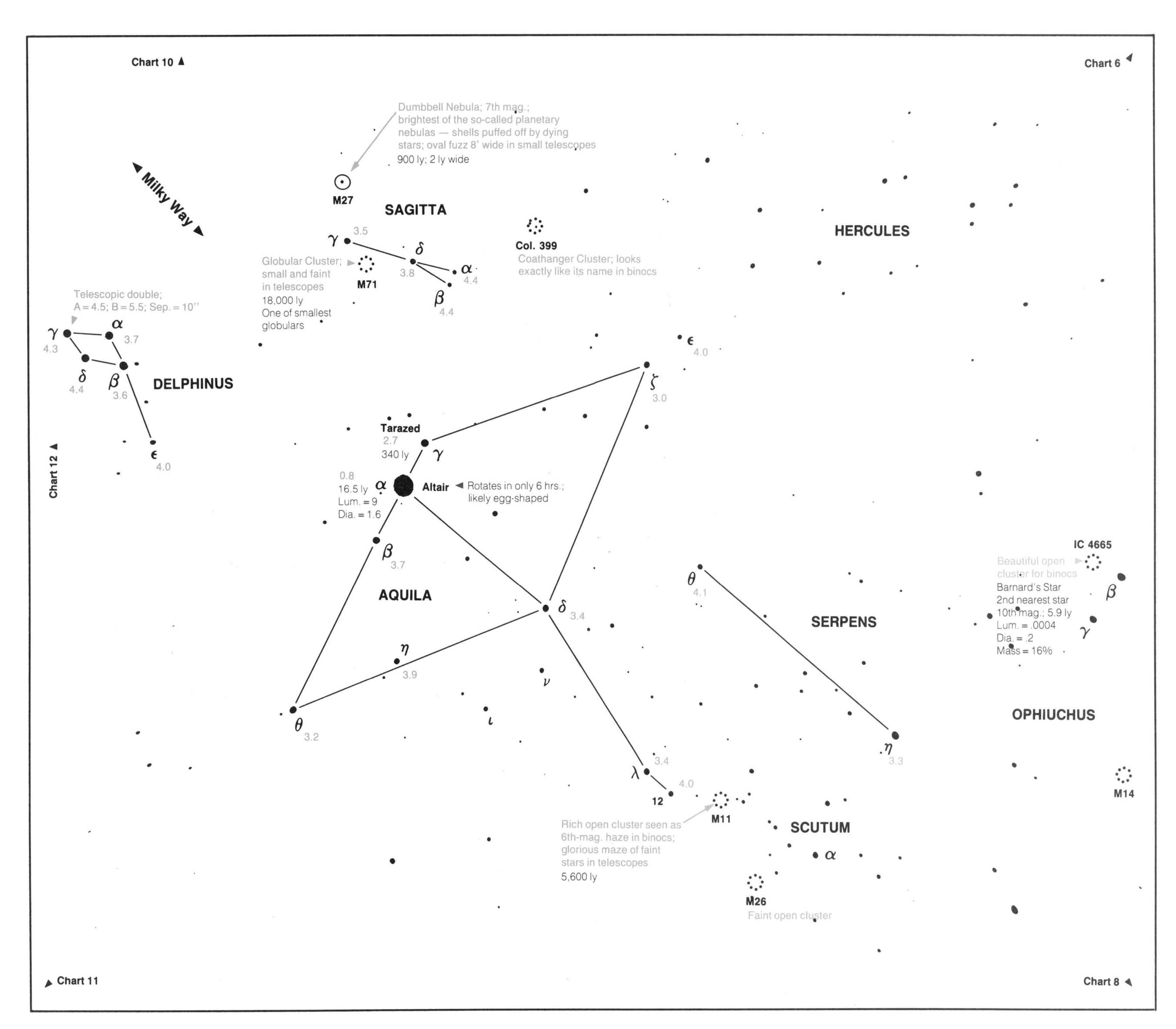

CHART 9

Aquila
Delphinus
Sagitta

Bottom part of Summer Triangle; seen throughout summer and early autumn.

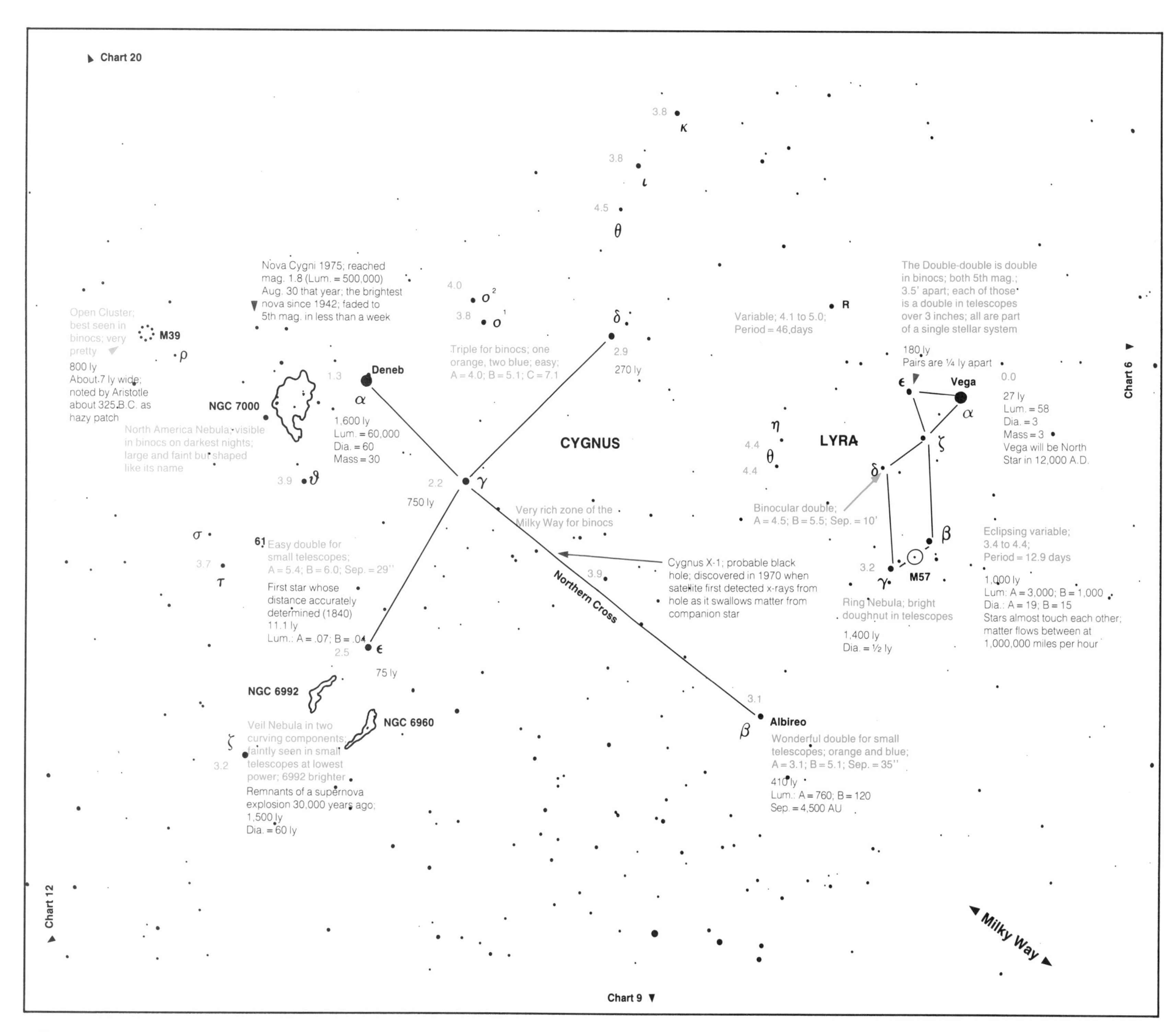

CHART 10

Cygnus
Lyra

Main section of Summer Triangle; in NE in spring, overhead in summer, in NW in autumn.

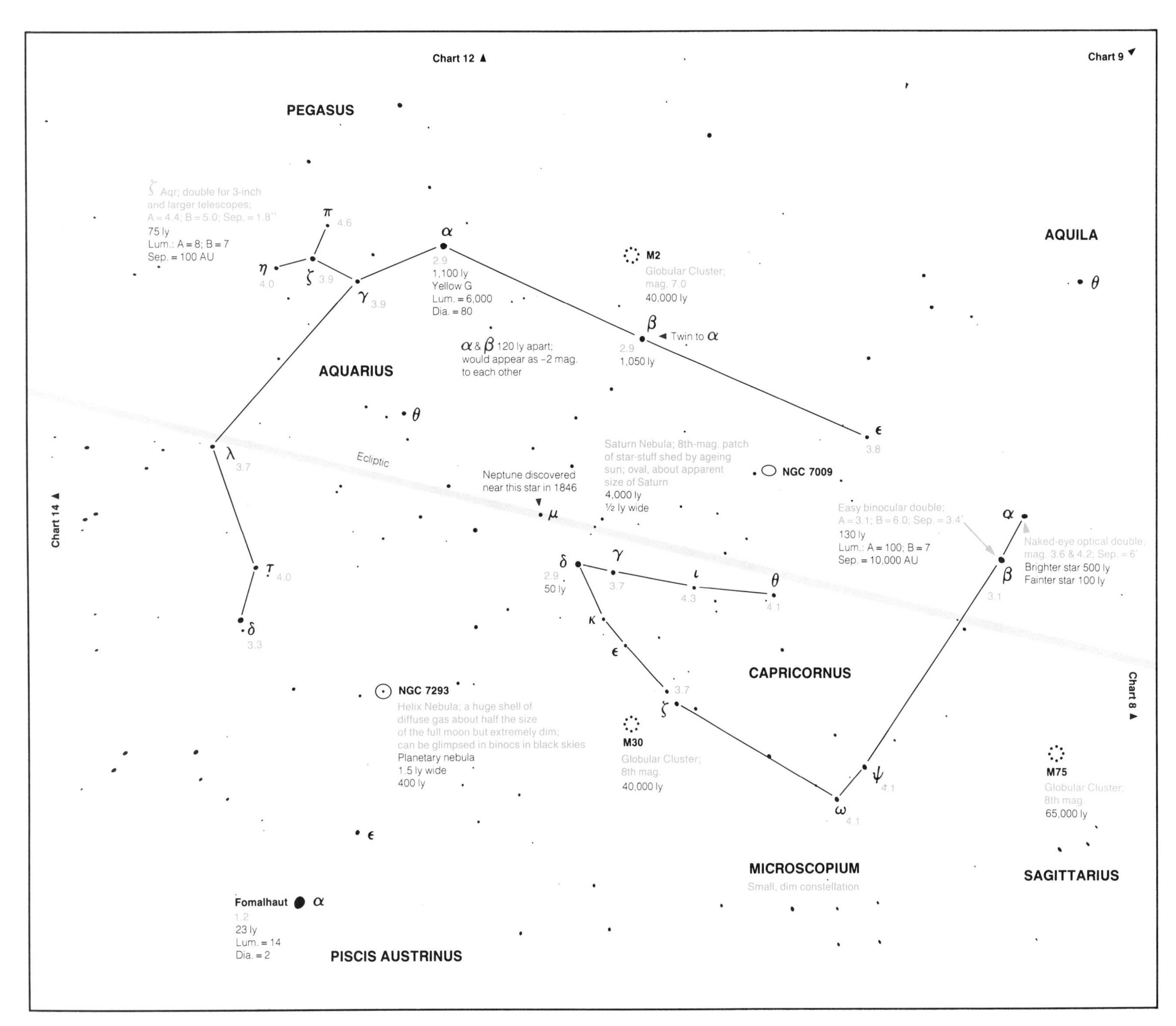

CHART 11 — **Capricornus Aquarius** — Two dim zodiac constellations seen in south in autumn.

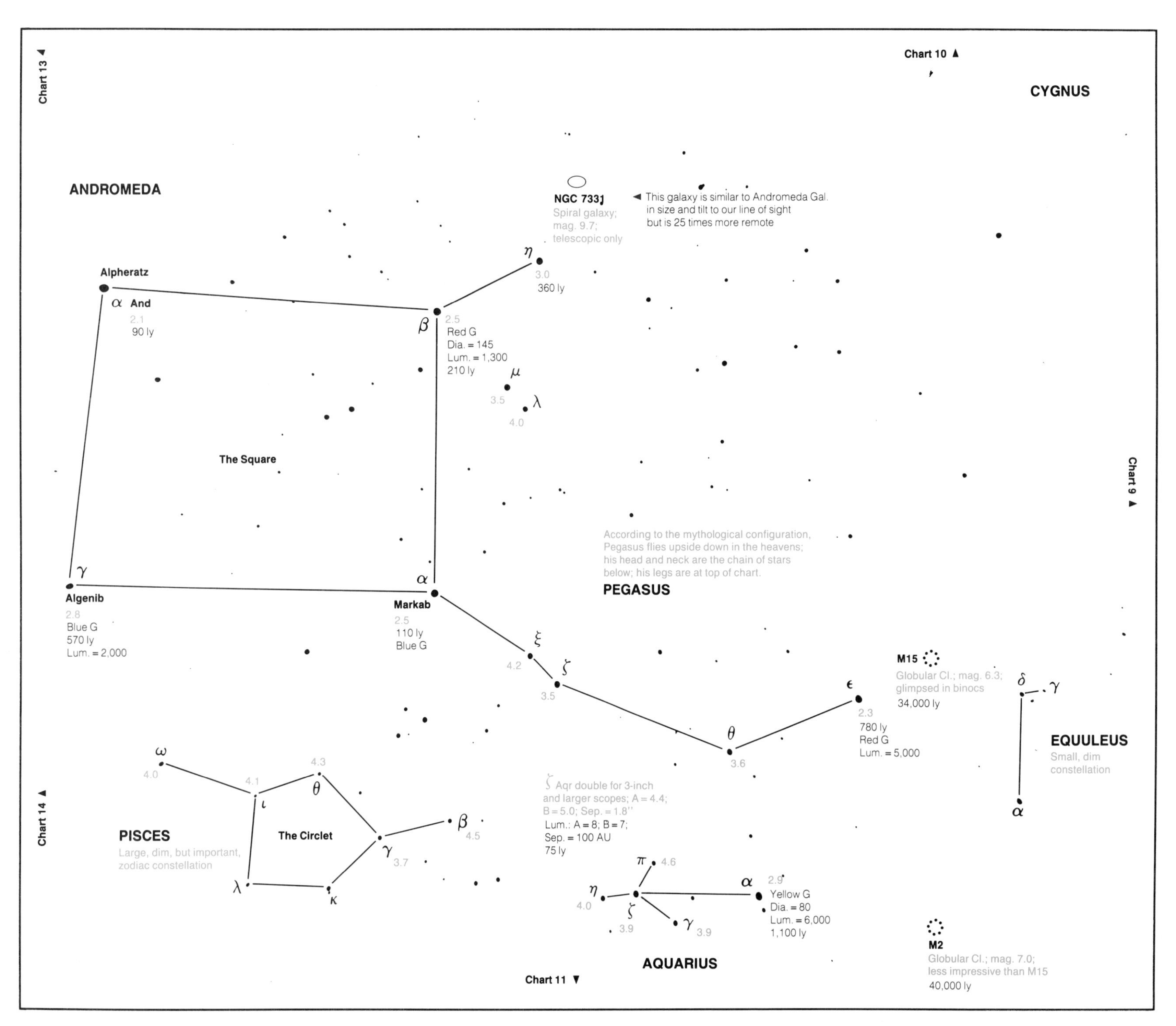

CHART 12

Pegasus
Pisces (part)
Aquarius (part)

A major guidepost region to autumn skies; near overhead October to December.

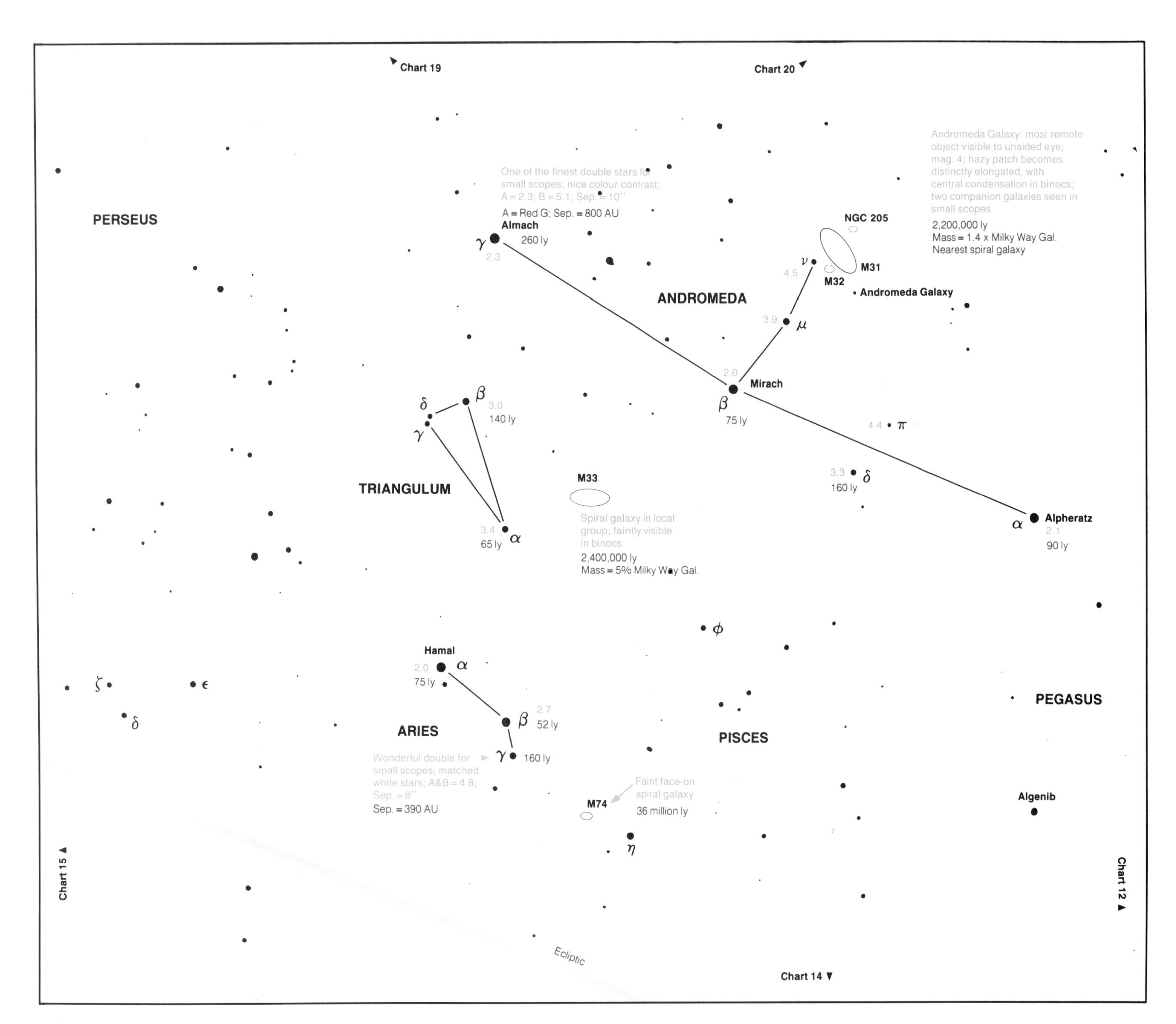

CHART 13

Andromeda
Aries
Triangulum

In NE sky in early autumn, overhead late autumn, in NW early winter.

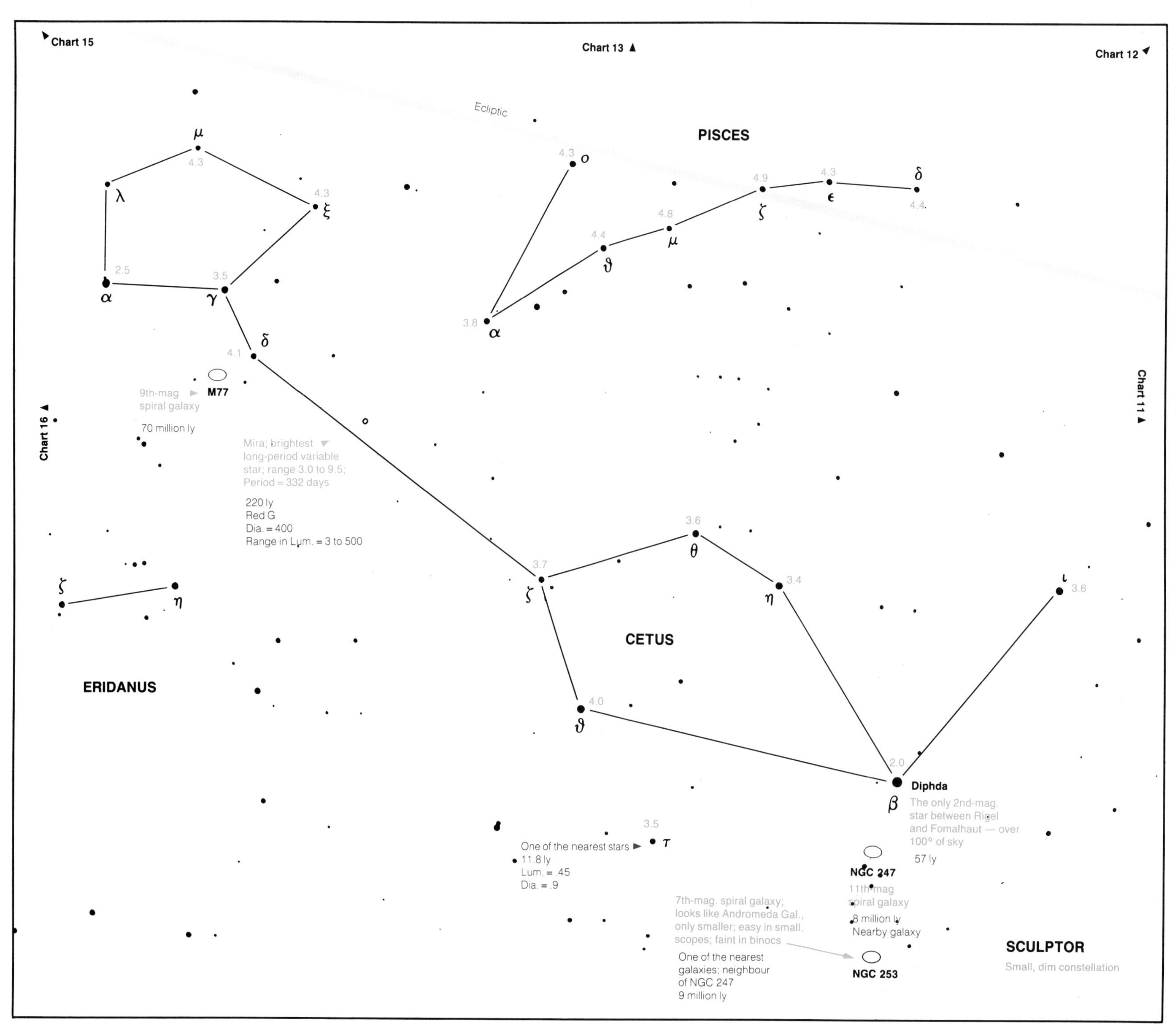

CHART 14 **Cetus Pisces** Dim constellations of the autumn southern sky.

Chart 15

Taurus
Auriga
Orion

Three of the six major constellations of the winter sky.

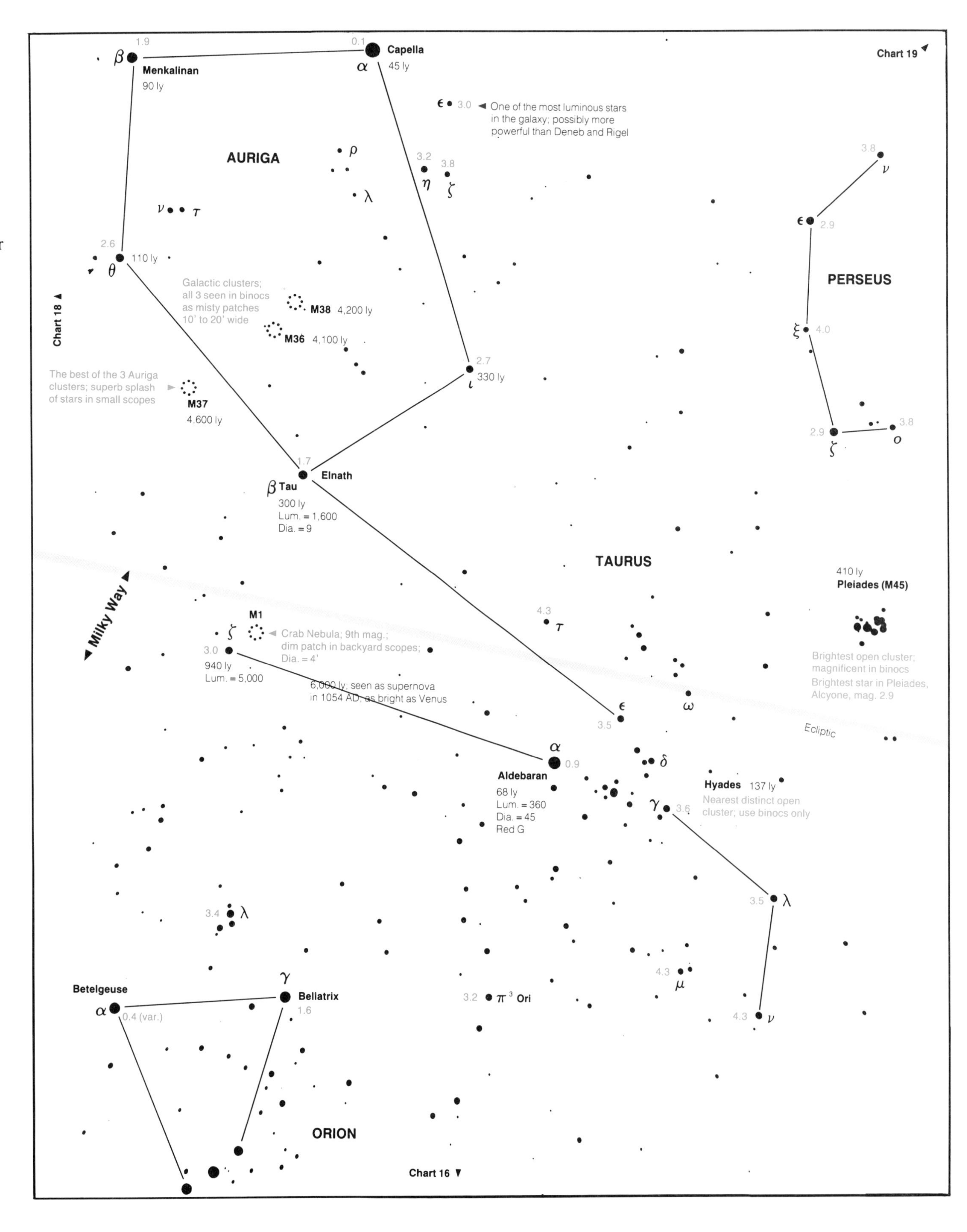

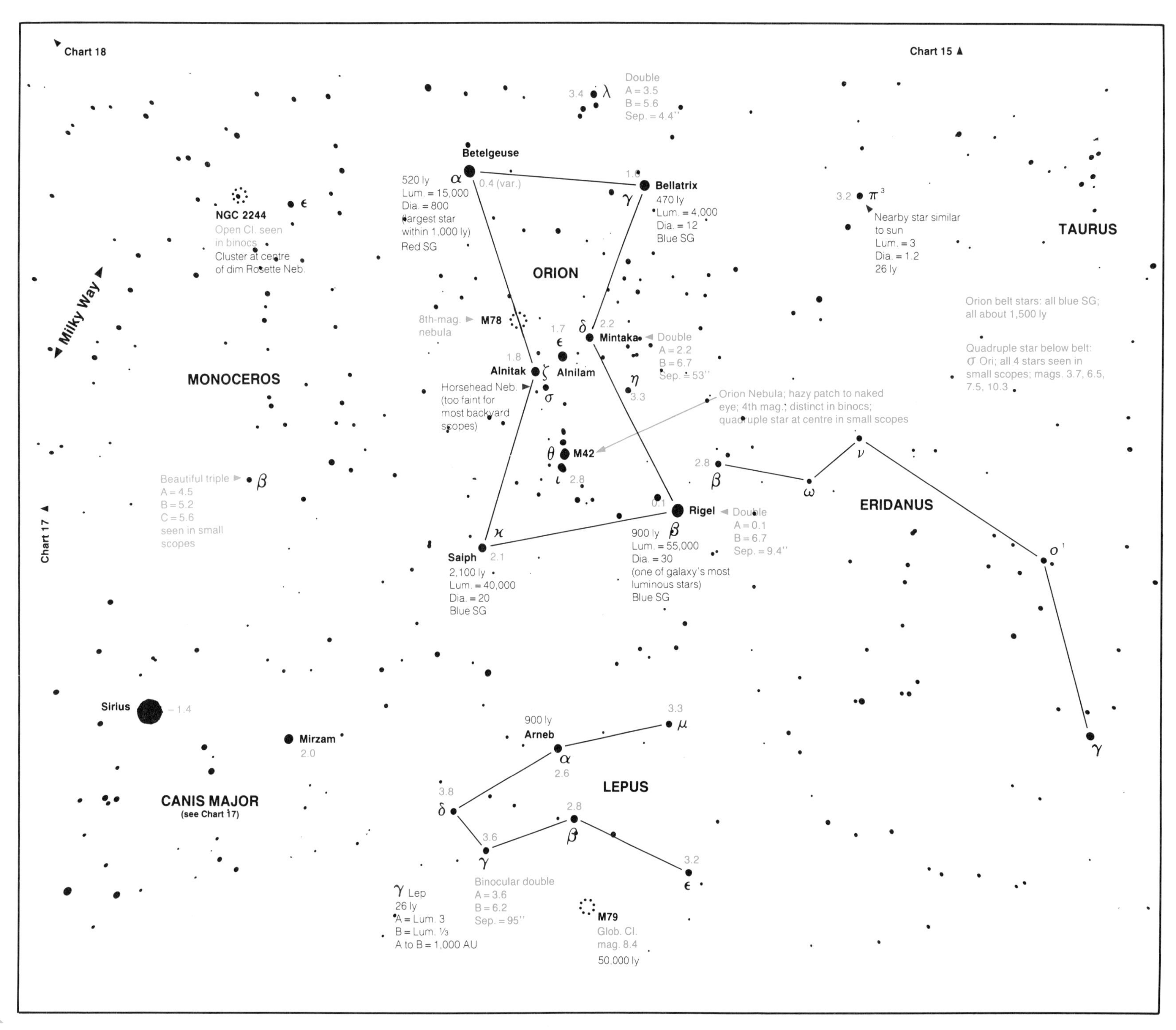

CHART 16

Orion
Lepus
Monoceros

Orion is the key to the winter sky, in addition to housing a vast array of celestial sights.

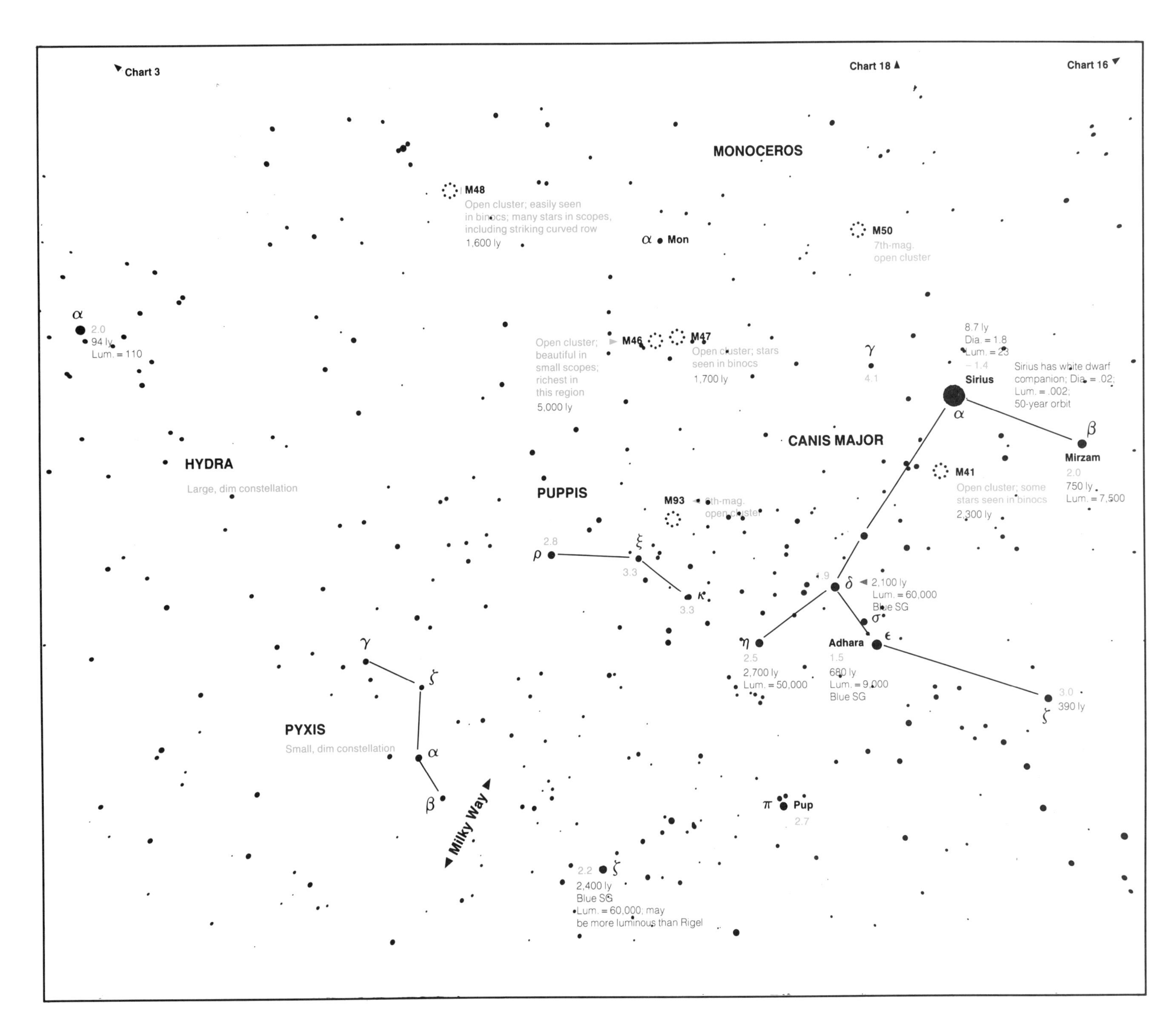

CHART 17

Canis Major
Puppis
Monoceros

Sirius, the night sky's brightest star, highlights this late-winter/early-spring sector of the southern sky.

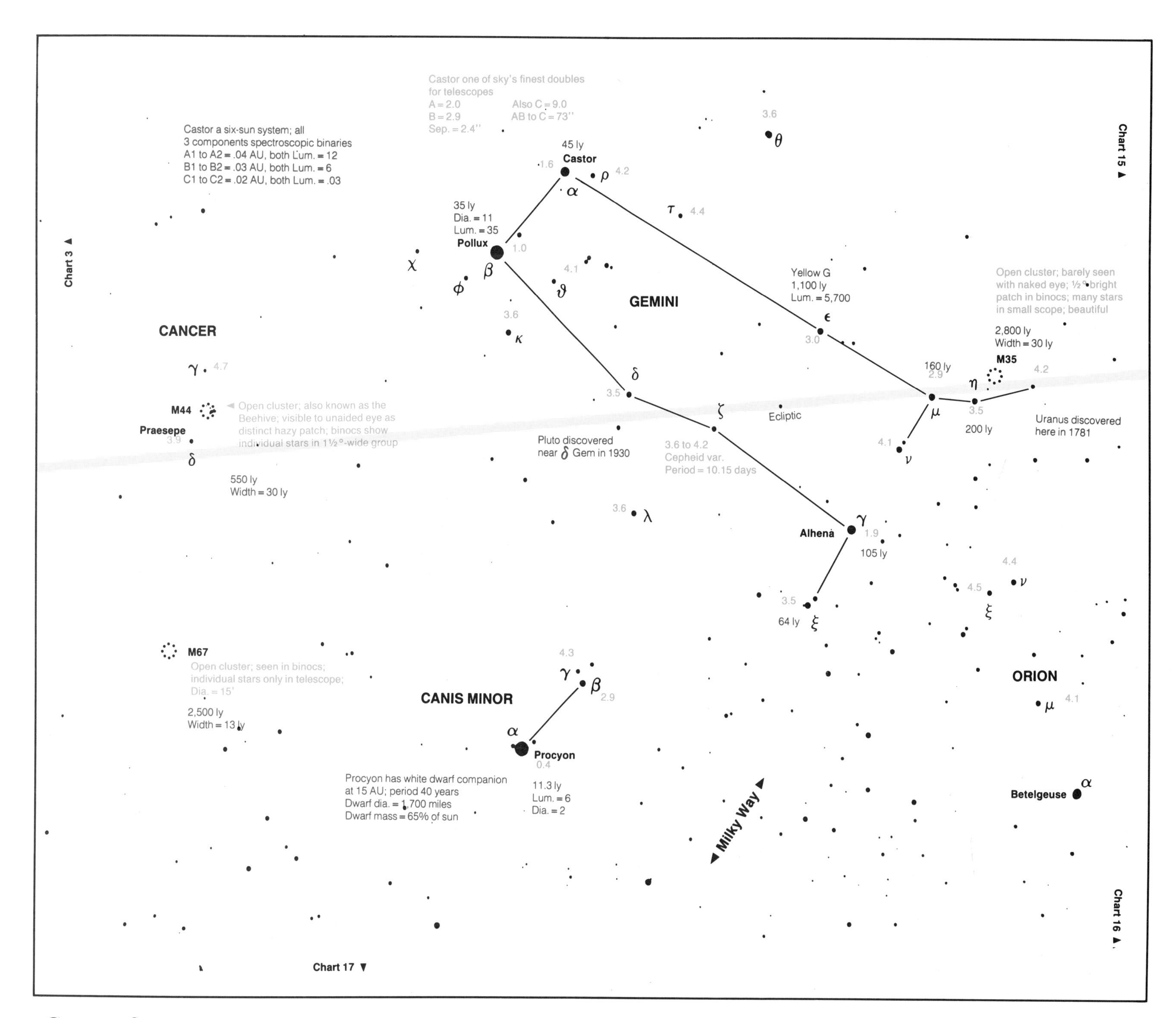

CHART 18

Gemini
Canis Minor
Cancer

This zone, east of Orion, is high in the south from midwinter to early spring.

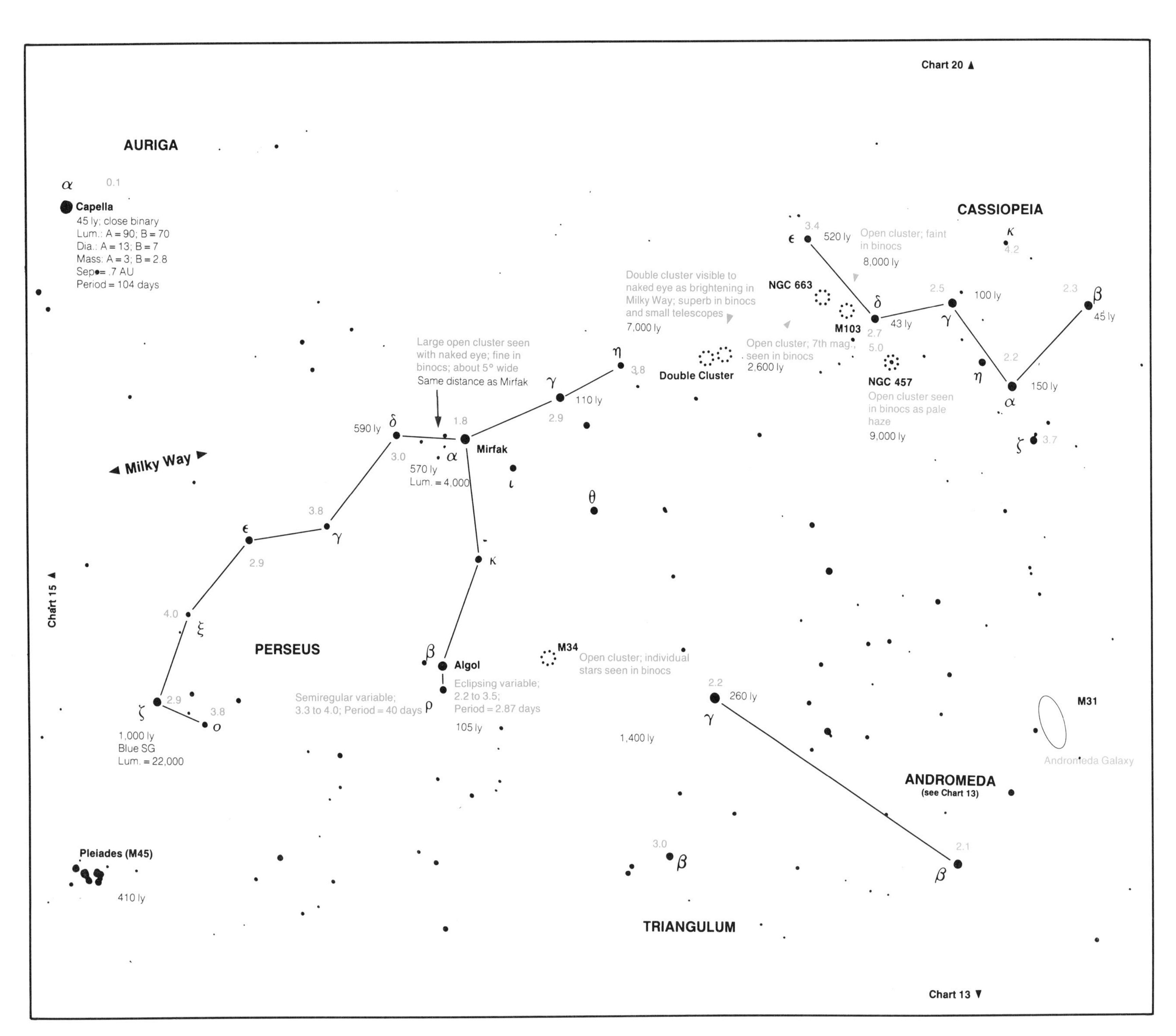

CHART 19

Perseus
Cassiopeia
Andromeda

In NE in autumn, overhead in winter, NW in spring.

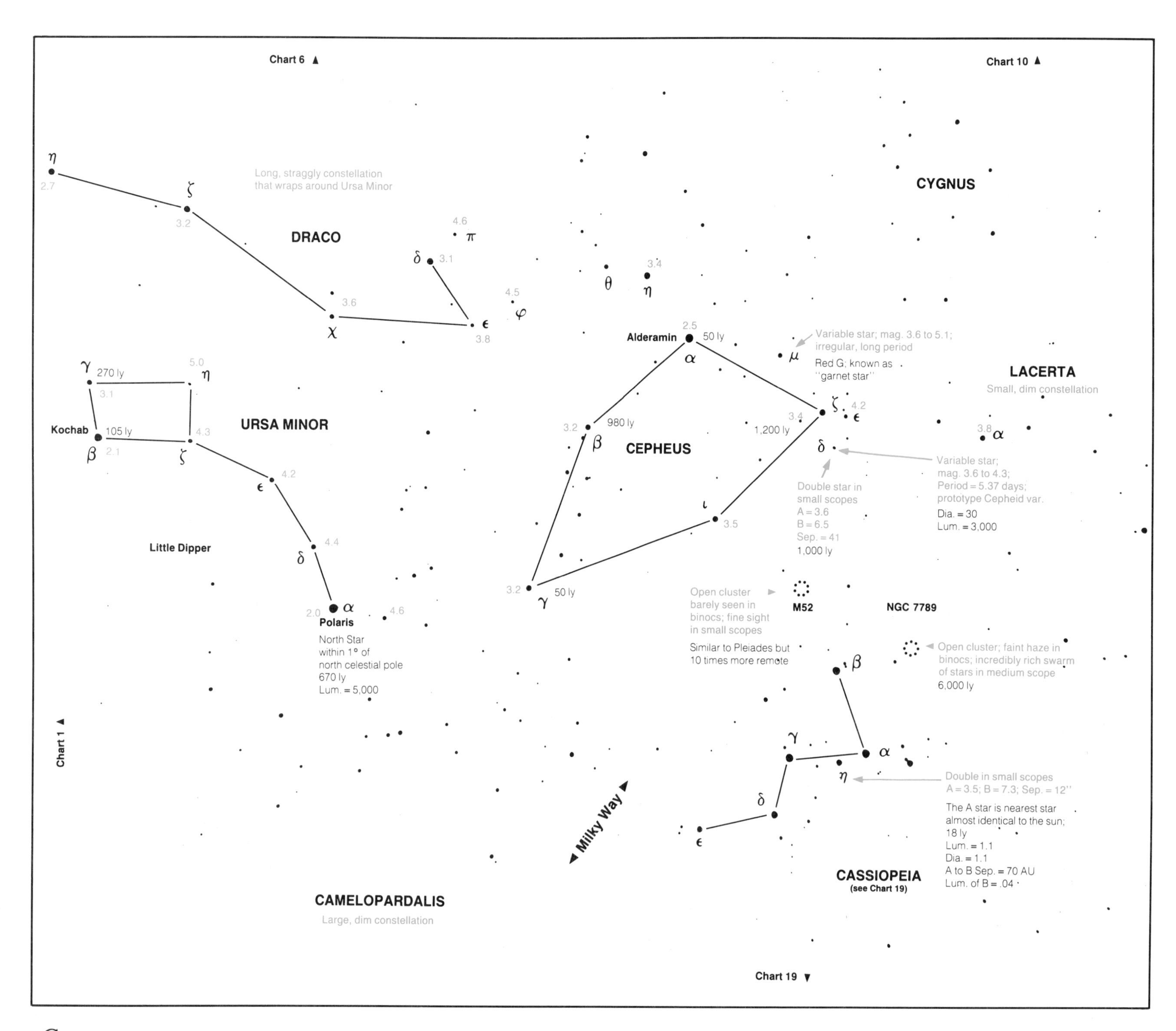

CHART 20

Cepheus
Cassiopeia
Ursa Minor

All of this chart except bottom third is visible all year; overhead in autumn.

7

The Planets

Now my own suspicion is that the universe is not only queerer than we suppose but queerer than we can suppose.

— J.B.S. Haldane

The oldest signs of culture on Earth — sketches on the walls of caves and markings on pieces of bone — reveal that as far back as 30,000 years ago, humans were trying to understand what they saw in the sky. Pondering the meaning of the sun, moon and stars must be as old as human inquiry. The existence of the constellations in lore and legend 5,000 years ago means that year after year, those ancient astronomers noticed the same stars and the same groupings. Among the celestial objects they observed were five bright wandering stars — the planets — objects that were, to our ancestors, propelled by some magical force.

These celestial rogues, masquerading as stars, roam across the sky. However, orbital geometry dictates that they remain confined to the zodiac constellations. Therefore, no planet ever swings by the belt of Orion or wrecks the configuration of the Big Dipper.

All the charts in this book have the ecliptic marked. If some bright object is observed in this region of the night sky, it is very likely a planet. Of the eight planets in the solar system besides Earth, five are visible to the unaided eye: Mercury, Venus, Mars, Jupiter and Saturn, all of them as bright as or brighter than first-magnitude stars. Of the remaining three, Uranus can be seen in binoculars, Neptune in a small telescope and Pluto in a large telescope.

Mercury, the planet nearest to the sun, is rarely seen. The planet's tight orbit keeps it so close to the sun that it pops into view for only a few weeks each year. When it is observed, the planet looks like a yellow zero-magnitude star huddled in the twilight glow after sundown or low in the east just before sunrise. Unless specifically looking for Mercury, a casual observer would probably never notice it.

Venus, the second planet out from the sun, and the one that comes nearest Earth, is the sky's premier jewel. It is so bright (magnitude −3 to −5, depending on where it is in its orbit) that those unfamiliar with the planet might think it is not a celestial object at all. Venus, dazzling white, is the first thing visible in the early-evening or early-morning sky for several months each year. Because Venus's orbit, like Mercury's, is between the sun and Earth, the planet is confined to a wedge on either side of the sun. It can never be seen for more than four hours after sunset or before sunrise.

Mars changes brightness far more than any other planet because its distance from Earth varies by a factor of 4, from 0.4 AU to 1.6 AU. Maximum brightness is magnitude −3, but that occurs only at the rare close approaches (1988, 2003). Normally, Mars is first or zero magnitude and shines with a distinctive rusty hue caused by sunlight reflected from its reddish deserts. Mars can move across more than half the sky during a year, making it the most interesting of all planets to watch as it tracks among the stars.

Jupiter is always brighter than any star but never as bright as Venus, its magnitude varying from −2 to −3. Shining with a yellowish white light, Jupiter is unmistakable. Like Mars and Saturn, it

The planet Venus, seen near the crescent moon, **above**, *is visible in full daylight if you look carefully. Mercury,* **right**, *hovering just above the horizon, is seldom as easy to see as it was in late April 1989 when this photograph was taken. See inside back cover for table of Mercury's visibility. Comparative sizes of the sun and the nine planets,* **left**, *clearly reveal two classes: small rocky worlds, like Earth, and massive gas giants led by Jupiter.* **Previous page:** *Composite of Voyager spacecraft images of Saturn and five of its satellites.*

can, at times, be seen all night stationed anywhere along the ecliptic because its orbit is beyond the Earth's. Jupiter spends about a year in each zodiac constellation, completing one trip around the sun in 12 Earth years.

Saturn is the planet most often mistaken for a star, since it equals the brightness of stars such as Regulus, Spica and Antares, rather than exceeding them as Jupiter and Venus do. And unlike Mars, Saturn does not have a distinctive colour, appearing simply as a pale yellow orb. Saturn requires 29½ years to complete its orbit and therefore spends at least two years in each zodiac constellation.

A long, steady look at a planet will usually reveal that it is not twinkling. Planets almost never twinkle like the stars, which invariably flicker, even on the stillest nights, due to atmospheric turbulence. The ubiquitous ripples in the Earth's air blanket easily disturb the pinpoint images of stars to generate the twinkling. Planets, on the other hand, are not pinpoints but tiny discs — too small for the eye to resolve but big enough for their light to be generally unaffected by the ripples, unless the atmosphere is abnormally agitated.

However, the surest way to identify planets is to know where they are relative to the constellations or, in the case of Venus and Mercury (which are usually seen before complete darkness), where and when to look. All of these facts are supplied in the tables on the inside back cover. The outer planets Uranus, Neptune and Pluto are below naked-eye visibility. To track these three planets with optical aid, use maps such as those in the *Observer's Handbook* or the *Astronomical Calendar* (see Chapter 12).

IS IT POSSIBLE TO SEE PLANETS OR STARS DURING THE DAY?

In general, the answer to this question is no. The only celestial object besides the sun that is normally visible during the day with the unaided eye is the moon, and even it is easy to overlook. But for several months each year, the next brightest object, Venus, is dazzling enough to be seen with the unaided eye in daylight.

Before Venus can be located in the daytime sky, a preliminary sighting in a dark evening sky is necessary. (See table to determine the dates when Venus is seen in the evening.) Watch for Venus as the sky slowly darkens after sunset. As soon as it is spotted, pinpoint its position by using a telephone pole, a tree branch, a chimney, a light standard — anything that projects into the sky. The next clear night, earlier in the evening, stand in the same location, and look for Venus slightly above and to the left of the marked position.

Venus appears to move about the width of a thumb held at arm's length (approximately two degrees) in eight minutes. This is not Venus's motion but the Earth's rotation. Therefore, if Venus was previously observed 15 minutes after sunset, search the sky right at sunset two outstretched thumb widths (roughly four degrees) above and to the left of the marked position. Keep backing up in this manner until Venus is viewed well before sunset.

Using this method, I have found Venus in a clear blue sky at least one hour before sunset. However, the sky must be deep blue; any haze greatly reduces the contrast between the planet and the sky and virtually rules out a daylight sighting. A more direct, though less challenging, method of daytime observation is to scan for Venus with binoculars. It is surprisingly bright in these instruments, but without some guidelines for where to look, it might require a fairly lengthy search.

Venus is usually 10 times brighter than the brightest star and 5 times brighter than Jupiter, the next brightest object in the sky. Although I have never succeeded in picking out Jupiter with the unaided eye before sunset, I have identified it with binoculars around sunset.

Sighting bright stars by day is possible when using telescopes equipped with accurate setting circles. I have found Vega, Sirius, Procyon and Altair this way, but it is really an academic exercise. There is a persistent legend that stars can be seen with the unaided eye in full daylight from the bottom of a dark shaft. The specific example cited in many books is the story that the star Thuban, which was once the pole star, was visible up a sighting hole constructed within the Great Pyramid of Cheops, from the pharaoh's burial chamber to the outer face. It is said that the star could be seen once each day from the pitch-black cavern within the pyramid. In 1964, two of these so-called sighting holes – long since filled with rubble and debris – were examined in detail by astronomers and Egyptologists. One hole did indeed point to Thuban, while the other aimed at Alnilam, the middle star in Orion's belt. However, the stars could not possibly be seen in full daylight. The shafts were likely symbolic of the pharaoh's celestial journeys to these two stars, each of which had significance in Egyptian religion.

To determine once and for all whether a star could be seen in daylight from the bottom of a dark shaft, a University of Cincinnati professor took his astronomy class into the base of an abandoned smokestack an hour before the star Vega passed precisely overhead. With their eyes fully dark-adapted in the gloom below, the students waited until the interval when Vega was aligned with the stack's opening. The appointed time came and went. They saw nothing, even though some resorted to using binoculars. The brightness of the sky was just too overwhelming.

ASTRONOMY FROM THE CITY

Planet observing with a pair of binoculars or a telescope is one of the few astronomical activities that can be conducted from the city almost as well as from dark rural locations. The planets are bright enough, except when near the horizon, to cut through city murk and pollution, providing accessible targets for urban astronomers. A telescope set up on an apartment balcony provides virtually the same views of Jupiter, Saturn, Venus and Mars as can be obtained at a farm or cottage. My 7-inch refractor, located in a backyard surrounded by streetlights and immersed in the glow of suburban Toronto, provided some of the best views of the planets that I have ever had. Seeing can be just as good (or bad) in the city as elsewhere. The glare washes out fainter objects, but the planets, being among the sky's most brilliant celestial lamps, are largely unaffected. The sun and moon (see Chapter 8) are even less obscured by light and haze, offering more targets for urban skywatchers.

MERCURY

Mercury, the planet nearest to the sun, is almost a twin of the Earth's moon. Both have heavily cratered surfaces that have remained essentially unaltered for the last three-quarters of the 4½ billion years since the solar system's birth. The craters are scars from what must have been ferocious bombardments by debris left over after the formation of the planets. The Earth's crust has undergone many changes since then, but the surfaces of Mercury and the moon have not.

It is as difficult to observe details on

Mercury as it would be to examine the surface of the moon if it were 300 times farther from Earth. A telescope magnifying 150 times (typical for planetary inspection) shows Mercury just half the size of the moon seen with the unaided eye. Added to that is Mercury's location near the sun's glare. For these reasons, only a few smudges have ever been recorded from Earth. Our knowledge of Mercury's surface stems entirely from photographs taken by the U.S. spacecraft Mariner 10, which flew past Mercury in 1974, the sole human artifact to visit that planet.

However, backyard astronomers can see the phases of Mercury. All but the smallest telescopes reveal a tiny crescent or half phase during the latter half of Mercury's prime evening visibility window each spring. The phases are simply the varying amounts of the day and night sides visible from Earth as Mercury swings in its orbit.

Because Mercury is easily identified only when it is close to the western horizon after sunset (or the eastern horizon before sunrise), horizon-induced poor seeing sometimes obliterates a clear view. Mercury is a small, scorched, dead world on the solar system's inner fringe — a challenge to the backyard astronomer. I find satisfaction in just locating the elusive planet. I feel fortunate if I get one sharp telescopic view of Mercury each year.

VENUS

For five or six months every year and a half, a brilliant object hovers over the southwestern horizon each clear evening after sunset. This is Venus, second only to the moon in brightness in the night sky. Venus's nearest rival, Jupiter, is never more than half as radiant, while Sirius, the brightest night star, is only eight percent as luminous.

Venus's dominance is largely due to its nearness to Earth. No other planet approaches as closely. Its minimum distance is 26 million miles, about 100 times the moon's distance. As the unsurpassed beacon in the night sky, Venus has figured prominently in the religions and cultures of many peoples. But no other elevated Venus to the level of importance it held for the Mayan civilization of Mexico.

At their peak, about 1,200 years ago, the Maya developed a sophisticated calendar system based on a 584-day period — the interval required for Venus to return to the same position in the sky. They called Venus the Ancient Star and seemed to be obsessed by its cycle of visibility.

A Spanish historian, Bernardino de Sahagún, studied the Maya in the 16th century and reported that enemies were sacrificed when Venus became visible after being lost for a few weeks in the sun's glare. "When Venus made its appearance in the east," he wrote, "[the Maya] sacrificed captives in its honour, offering blood, flipping it with their fingers toward the planet."

Another reason for Venus's dazzling appearance is a cloak of brilliant white clouds that reflect 72 percent of the sun's light back into space. At only two-thirds the Earth's distance from the sun, Venus receives twice as much sunlight.

Venus is regarded by many scientists as an Earth gone wrong. The planet is almost identical in size and mass to Earth, yet it is shrouded in an atmospheric cloak 90 times as dense. This blanket has turned Venus into a cosmic greenhouse, effectively trapping the solar radiation that penetrates the clouds. The surface temperature is a constant 400 degrees C.

The Venusian atmosphere is almost entirely carbon dioxide laced with sulphuric-acid droplets. It is inconceivable that any form of life could have evolved on Venus.

The surface of Mars, **above,** *has proved to be a dust-blown boulder-strewn landscape — far different from visions of canals and vegetation proposed earlier this century. Fine particles of dust lifted from the surface during windstorms remain suspended in the carbon-dioxide atmosphere for years, producing the peach-coloured sky. The temperature at the time the Viking 1 lander took this photograph was about minus 35 degrees C. Some of Viking's weather instruments are located on the spacecraft boom visible at centre.* **Left,** *author's sketch of Jupiter, as seen through a 3-inch refractor in 1962. Black dot is the shadow of the Jovian moon Ganymede.* **Far left:** *During a dawn stroll on the catwalk of the Canada-France-Hawaii Telescope at an altitude of 13,800 feet, the author captured a solar system traffic jam. The three brightest objects, apart from the crescent moon, are (top to bottom) Jupiter, Venus and Mercury.*

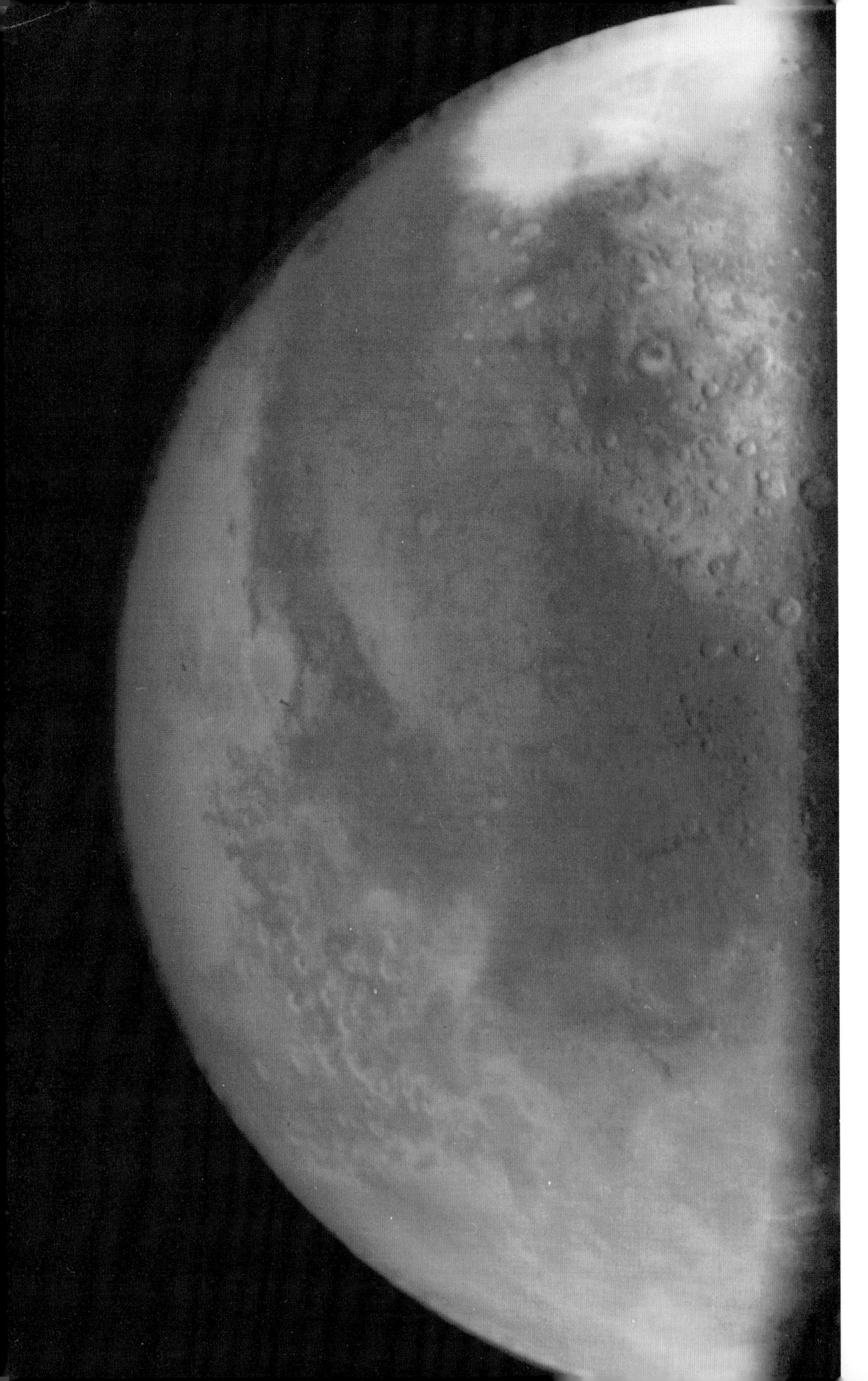

Even exotic science fiction scenarios for silicon-based and other life forms seem unrealistic in the Venusian environment.

I like to observe Venus telescopically just around sunset, when it looks like a cue ball suspended in the deepening blue. The snow-white surface — the cloud blanket — is totally featureless, its blank stare an appropriate mask for the furnace-hot wasteland below.

Like Mercury, Venus cycles through phases from nearly full to a thin crescent, but its phases are much easier to detect. A good telescope will produce a sharply defined image of the planet, large and dramatic when Venus nears its closest point to Earth once every 1½ years. At such times, it is a beautiful sickle-shaped crescent.

But in general, Venus is not a compelling telescopic subject, since nothing apart from the phases can be seen. When it decorates the evening sky, Venus is usually the first object I turn my telescope toward. But after a few minutes, I begin tracking down more varied targets. The queen of the night is never unveiled to backyard astronomers.

MARS

The old Mars, the Mars that we knew before the space age, was a wonderful world of heroes, maidens, bizarre creatures and — most captivating of all — a dying civilization desperately attempting to prolong its existence by constructing a global network of canals to preserve dwindling water supplies. Peering from Earth, turn-of-the-century astronomers saw those canals, or thought they did.

The craters, polar ice caps and reddish landscape of Mars, **left**, *as seen by the Viking 1 orbiter in 1976. The same colours are visible in backyard telescopes, although the amount of detectable detail is, of course, much less. The three photographs,* **right**, *taken during the 1988 close approach of Mars using a 23-inch refractor telescope, show what can be seen under optimum conditions with a 4-to-8-inch instrument. South polar cap is prominent.*

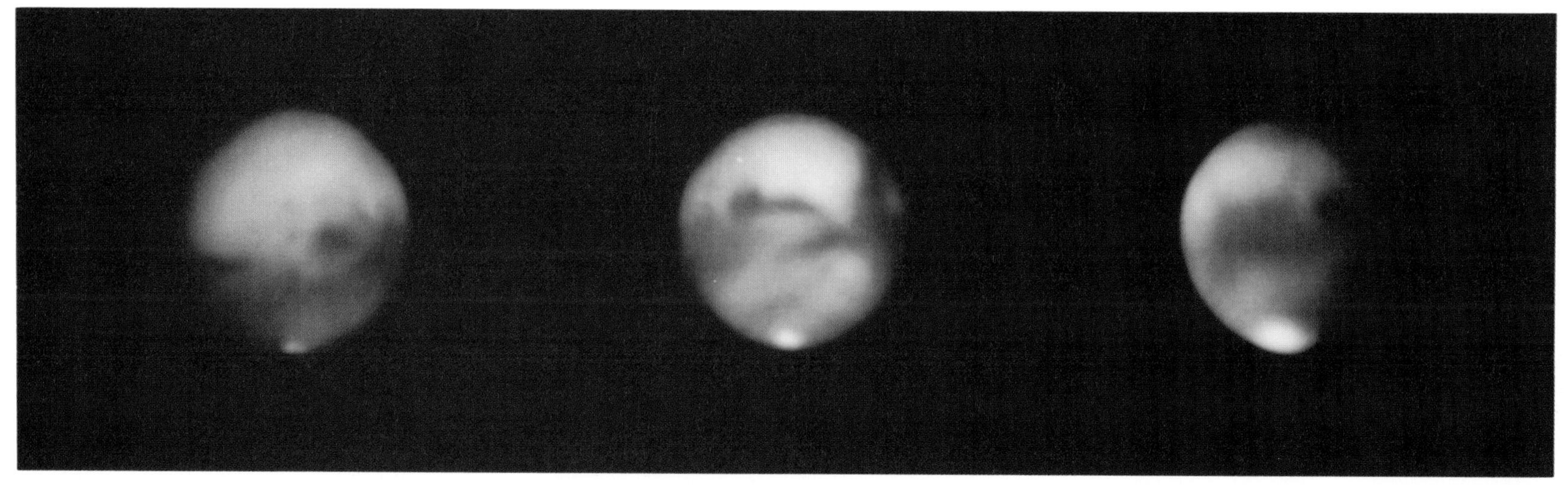

Today, the mystery of the famous canals is gone. They proved to be not waterways but optical illusions born in the minds of observers who unconsciously linked subtle detail near the threshold of vision into linear features.

Speculation about Martian civilizations has given way to the real Mars, a planet midway between Earth and the moon in size and surface conditions. Spacecraft images of deserts, craters and colossal volcanoes that dwarf Mount Everest revealed a world less Earthlike than almost anyone suspected. Soon after the sophisticated life-searching equipment on board the American Viking 1 lander turned in negative results, plans for follow-up missions to Mars by both the Americans and the Russians were postponed for more than a decade. The canals and Martians faded away, replaced by hard reality.

I vividly remember the final transition from the old Mars to the new. It was a warm summer evening in July 1976 when the first Viking touched down on Mars. I was at "mission control" at the Jet Propulsion Laboratory in Pasadena, California, with about 200 scientists and an equal number of journalists and science fiction writers anxiously awaiting the first surface photographs to be beamed from Mars.

The images began to build up line by line on the television monitors, and a boulder-strewn landscape of sand dunes finally emerged. Science fiction writer Ray Bradbury, who happened to be standing beside me, said quietly: "From this moment on, we don't have to imagine what Mars is like anymore."

And so it has been with most of the other major worlds in the solar system. They have been removed from the borderline of science and science fiction and brought sharply into focus by the electronic eyes of robot spacecraft that Earthlings have flung to all but the outermost planet of the solar system.

Parallelling this, the focus of backyard astronomy has changed. Prior to the space age, planetary observation was the chief activity because so little was known about our neighbouring worlds. The aura of the unknown was a powerful magnet keeping hobbyists hunched over their eyepieces for hours to catch a glimpse of a Martian canal or an alteration in the red planet's dark zones (believed to be vegetation as recently as 1965 by a few experts).

Today, the emphasis has shifted. Examination of Mars through the telescope is more casual, less encumbered by inquiry – more like a tourist's view of a well-known vacation island from an aircraft. Interesting, somewhat exotic, but not mysterious. When I view the Martian disc now, I think of the winds that whip the dunes of the vast deserts and howl down the great Mariner Valley, a canyon five times deeper and a hundred times longer than the Grand Canyon. I see from afar the endless plains painted red by layers of iron-oxide minerals that give the planet its unique rusty-coloured hue and the vast polar ice caps trimmed by a sheet of solidified carbon dioxide – part of the planet's thin atmosphere, frozen to the ground over the winter.

Yet, as it has for millennia, the red "star" dominating the summer evening sky still attracts attention. Every 26 months or so, Earth catches up to Mars and overtakes it. The point of passing, or closest approach, is called opposition. Because Mars has an elliptical orbit, the distance between the two planets can vary from 0.38 AU to 0.68 AU from one opposition to another. However, even the greater distance is considered close by planetary standards.

The problem for backyard astronomers is that Mars is a small planet, only twice the size of the moon. Even at its absolute minimum distance, Mars appears barely larger than the globe of Saturn and nowhere near as big as Jupiter. Sometimes, when Mars and Earth swing to opposite sides of their orbits, the red planet shrinks to the same apparent size as Uranus. So it is not surprising that, as often as not, Mars is just a pale, rusty dot in the telescope's field of view.

Within a few weeks of opposition, however, Mars does reveal itself to the backyard observer. Because of its tenuous atmosphere (0.7 percent as dense as the Earth's), Mars is the only planet whose surface features are clearly visible from Earth. But those features are fairly subtle, except for major dark areas like Syrtis Major or the brilliant white polar caps.

When I first looked at Mars through

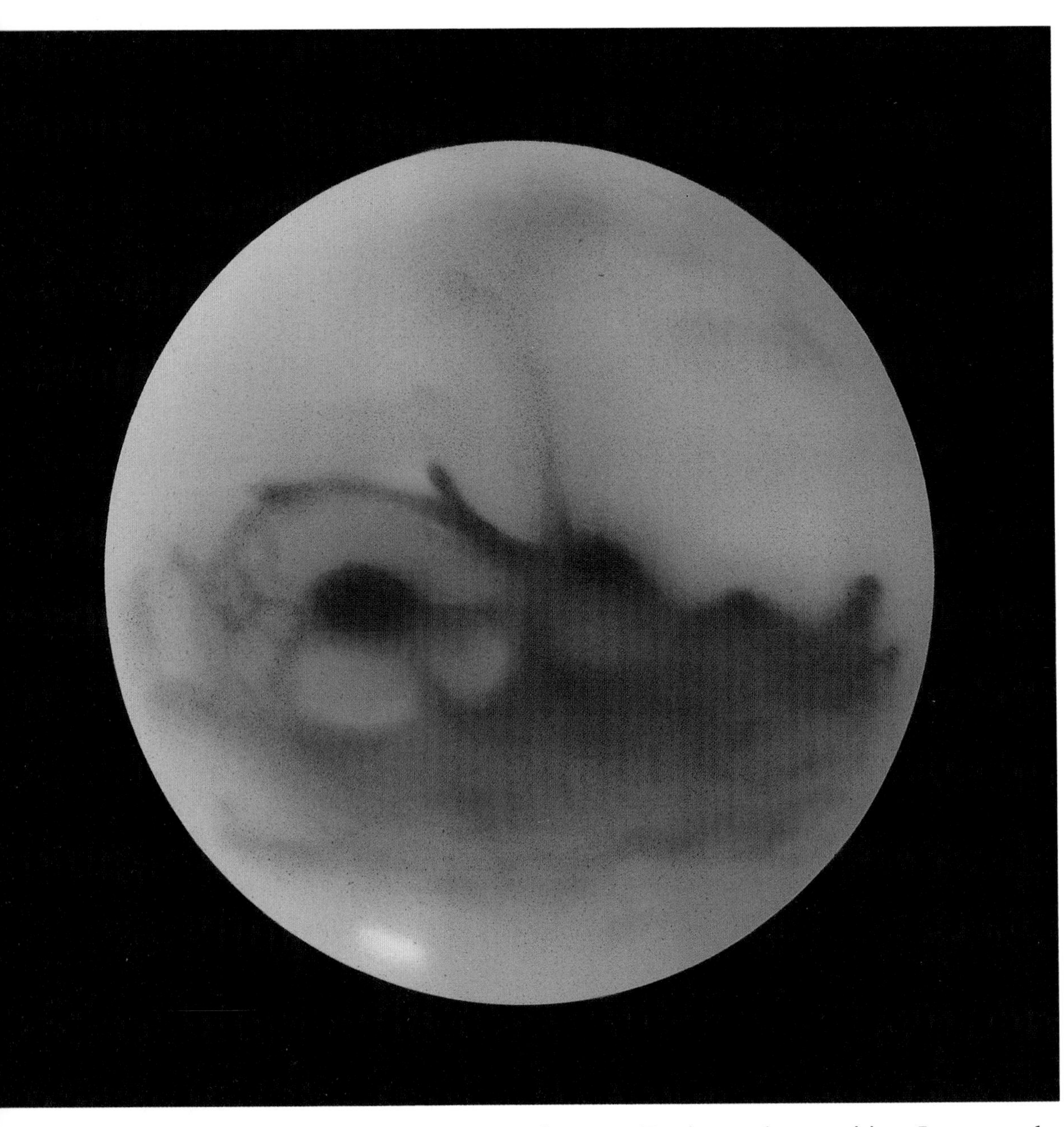

my 3-inch refractor, I could discern nothing more than the white button of one of the polar caps. It was weeks before I was able to train my eye to see more. Even now, every other year when Mars returns, I have to educate my eye again, although it takes only a night or two. Telescope, mind and eye work together to reveal more and more each night. Since Mars's rotation is only 40 minutes slower than the Earth's, the same face is seen 40 minutes later the following evening, which aids in the familiarization process. Phobos and Deimos, the two moons of Mars, are much too small for detection in amateur telescopes.

During each opposition, I try to make enough sketches of Mars to produce a small map of the planet. I use a two-inch circle and a soft pencil, and I keep detailed notes of the intensity of the features, indicating whether they were easy or difficult to see. I record the time and the date so that I can calculate the planet's central meridian longitude from tables in the *Observer's Handbook*. Then I can plot what I saw on my own map of Mars.

In past decades, amateur astronomers made thousands of drawings of Mars, some showing more detail than the best observatory photographs. (Unlike long-exposure photographs of sky regions that contain nebulas and galaxies, short exposures of the planets are always blurred somewhat by seeing variations which affect larger telescopes.) At one time, amateur observations were coordinated by international astronomical societies as a means of monitoring changes in the appearance of the quasi-permanent surface features. Such programmes were largely abandoned when spacecraft began photographing the planet from orbit in 1971.

But using scientifically valuable results as a justification for observing Mars or for any other aspect of backyard astronomy is an argument that has never impressed me. Not that I haven't reported some of my observations and had them published in respected journals. If the observations are made, they should be reported. However, the vast majority of backyard astronomers are out there exploring the celestial sights for one purpose only — enjoyment and personal edification. And that is reason enough.

The thrill of seeing Syrtis Major or of catching a fleeting glimpse of the Chryse Plain, where Viking 1 landed, are the rewards. Knowing that the great Nix Olympica volcano looms between Tharis and Amazonis stokes the imagination. The new knowledge of Mars, revealed by the television eyes of Mariner and Viking space probes, has shown that planet to be an exciting, varied world which remained unrecognized during centuries of Earth-bound investigations.

THE ASTEROID BELT

The popularized image of asteroid belts as outer-space pinball machines, with craggy crater-pitted boulders practically bumping into each other, makes exciting science fiction, but it departs from reality.

The zone occupied by the asteroid belt, between the orbits of Mars and Jupiter, is enormous. Even though there are more than a billion asteroids the size of a house or larger, the space they occupy is so vast that an astronaut standing on one of these rocky bodies would only rarely see another asteroid appearing brighter than a third- or fourth-magnitude star. This situation became vividly apparent when the Voyager-spacecraft trajectories were being planned in the early 1970s. Analysts searched in vain for a flight plan that would meet the mission requirements at Jupiter and Saturn and would also come close enough to an asteroid — any asteroid — to allow Voyager's high-resolution cameras to have a peek. Far from being a hailstorm of solar system gravel, the asteroid belt is essentially as free of debris as the region around Earth.

Asteroids were unknown before the 19th century — but not entirely unsuspected. Astronomers were puzzled by the wide, apparently vacant, zone between the orbits of Mars and Jupiter that spoiled the otherwise regular progression of worlds outward from the sun. So strong was the feeling that something must occupy this region that several astronomers spent years searching for the "missing" planet.

On the last evening of the 18th century, Giuseppe Piazzi, an Italian monk who was revising a star atlas, noticed a star that was not plotted. He checked its position the following night and found that it had shifted slightly. When the new body's orbit was calculated, astronomers realized that it wheeled around the sun right in the middle of the Mars-Jupiter gap. The object, named Ceres, was apparently much smaller than the moon.

Skywatchers never had a chance to get used to the idea of one small planet between Mars and Jupiter because a second one, Pallas, was soon discovered. Two more tiny planets, Juno and Vesta, were tracked down during the following years.

Today, more than 3,000 asteroids have been observed long enough for a determination of their precise orbital paths. The number of asteroids a half-mile in diameter or larger is estimated to be half a million. Undoubtedly, there are millions more, ranging from huge boulders to pebbles. Four are larger than one-tenth the diameter of the moon. Ceres, the biggest by far, at 580 miles in diameter, hoards one-third of the estimated mass of the rest of the asteroids combined. Next are Vesta (325 miles), Pallas (320) and Hygeia (255).

Keeping a diary of observations and sketching what you see, **above**, *is the best way to improve your observing skills. Drawing of Mars,* **left**, *is from the author's observations with a 7-inch apochromatic refractor in 1988.*

Yet of all these asteroids, only Vesta is occasionally barely visible to the unaided eye. However, dozens of asteroids can be seen as starlike objects in backyard telescopes. Sky charts of their positions are published in the *Observer's Handbook*, the *Astronomical Calendar* and *Sky & Telescope* magazine.

The existence of an asteroid belt instead of a planet can mean one of two things: that a planet was unable to form and the debris is evidence of the aborted process or that the planet did form but somehow shattered. The exploded-planet hypothesis has relatively few adherents. Most astronomers support the theory that proposes an original family of smaller objects which never coagulated into a major planet due to the disruptive gravitational influence of nearby Jupiter, the most massive planet in the solar system.

JUPITER

In one of Arthur C. Clarke's science fiction novels, an interplanetary spaceship approaches Jupiter from just beyond the orbit of Callisto, outermost of the planet's four large satellites. Here, more than one million miles from the solar system's largest planet, one of the crew gazes out the ship's window toward the colossal globe of Jupiter, resembling a multicoloured beach ball suspended in the sky by an invisible string.

As he watches, mesmerized by the quilt of storm-riddled clouds which swirl across the face of Jupiter, he notices that the great globe is spinning rapidly. Clouds that only a few hours ago were at the centre of the planet are now moving out of sight as new ones sweep into view. Clarke goes on to explain that the planet is not a world of rock or ice but a ball of liquid and gaseous hydrogen, the largest of the solar system's four gas giants.

What intrigues me about this is that Clarke's description of the view is not much different from what I have seen when examining Jupiter through my 7-inch telescope with an eyepiece magnifying about 180 times. A cooperative atmosphere that allows a distortion-free view of Jupiter puts me, in effect, just a few hundred thousand miles beyond Callisto. It is the next best thing to the view that I would have out the window of Clarke's interplanetary spaceship.

Binoculars are all that is needed to begin observing the Jovian system. They reveal the movements of the four large moons that orbit Jupiter in periods ranging from just under two days for Io (the nearest of the big four) to about 17 days for Callisto.

Steady the binoculars by resting them on a broom handle or fence, since your arms alone cannot hold them still enough for astronomical observations. For the best results, support them rigidly by using a tripod adapter (available from well-stocked camera stores) that clamps the binoculars to a camera tripod.

The moons of Jupiter provide an ever-changing "solar system" in miniature as they shuttle back and forth from side to side. Occasionally, the satellites are paired on either side of the belted gas planet or are all lined up on one side. Some of the moons may be lost in Jupiter's glare, or they may be behind or in front of the planet. Even in large telescopes, only the four largest of Jupiter's 16 moons are visible. The other satellites, all less than 200 miles in diameter, have been recorded photographically.

The major satellites in order from Jupiter are Io, Europa, Ganymede and Callisto. At magnitude 4.6, Ganymede, the largest, is theoretically bright enough to be visible without optical aid. However, Jupiter's brilliance prevents such observation — at least that has been the accepted wisdom on the subject. Yet for many years, there have been persistent reports by people who claim to have seen Ganymede with the naked eye. Some say that they have also detected Callisto, the most distant of the large moons, even though it is only magnitude 5.6, near the threshold of vision. I cannot see either of them myself, and I have always concluded that these sightings are biased because the observers know the satellites are there — a mind-eye combination that gives the illusion of a small dot beside the brilliant luminary.

After many years of inconclusive debate on this subject, it appears that at least some of these sightings are accurate. According to Xi Ze-Zong of the Chinese Academy of Science, Gan De — one of China's earliest astronomers — left records of observations of one of Jupiter's moons in 364 B.C. Gan De wrote: "Jupiter was very large and bright; apparently, there was a small reddish star appended to its side." Xi Ze-Zong says this passage almost certainly means that Gan De saw a Jovian satellite.

To test the validity of Gan De's statement, an experiment was conducted in a Chinese planetarium, which revealed that people with good eyesight can see a satellite the same apparent brightness as that of Ganymede and Callisto at their farthest orbital positions from Jupiter. The tests also showed that it would be even easier to see a blended image of the two. This new evidence suggests that Galileo's discovery of Jupiter's moons in 1610 was preempted by almost 2,000 years.

Tracking a Jovian moon as it casts its shadow on the disc of the giant planet is one of the great telescopic sights in astronomy. According to some observers, this phenomenon is visible in a good 60mm (2.4-inch) refractor at 70x to 100x. The tiny ink-black shadow of a moon can be spotted when it is on one of the brighter zones of Jupiter's clouds.

Because the satellites range in size from Europa, slightly smaller than our moon, to Ganymede, 1½ times the moon's diameter, and because of their varying distances from Jupiter, their shadows are of different sizes. Io and Ganymede produce shadows about 1,700 miles in diameter, whereas Europa and Callisto cast significantly smaller shadows, approximately 1,000 miles across. Sharp-eyed observers have seen the shadows of all four with 3-inch refractors, but for general conditions and the average eye, a 4-inch or larger telescope is required.

Both the satellites' orbital positions and the times when their shadows are visible on the planet's disc are given in the *Observer's Handbook* for the period when Jupiter is conveniently positioned in the sky for viewing. Also given are the times when each moon dips into Jupiter's shadow or disappears behind the planet. The moons' orbits are exactly in the plane of Jupiter's equator. But since Jupiter stands up almost vertically, with its axis tipped only three degrees, the plane of the equator and the orbits of all the satellites are seen nearly edge-on. It is like watching a billiard game with your eyes at table level. The moons swing back and forth without deviating from a thin zone on either side of Jupiter.

The windswept cloud deck of Jupiter is continually changing, the vast dark belts merging with one another or fading to insignificance. The bright zones — actually smeared bands of ammonia clouds — vary in intensity and are frequently carved up with dark rifts or loops, called festoons. The most dramatic action occurs in the equatorial zone.

The clouds at Jupiter's equator rotate five minutes faster than those on the rest of the planet: 9 hours 50 minutes compared with 9 hours 55 minutes. This means a constant atmospheric-current interaction as one region slips by the other at approximately 250 miles per hour. Besides these disturbances, there are changes in the intensity of the various belts and zones from year to year.

Despite its great distance, Jupiter appears far larger in the telescope than any other planet. Even when seen at their best, all of the other planets combined do not equal Jupiter's visible surface area. At least two of the equatorial cloud belts will be seen in any decent telescope, and a quality instrument will reveal a multicoloured globe of atmospheric zones.

Jupiter's rapid rotation has caused the great globe to become markedly oval so that it appears to be about seven percent squashed at the poles. The variety of cloud features and the choreography of its four large moons make Jupiter one of the top backyard attractions for amateur astronomers.

SATURN

Saturn is the superstar of the night sky, a dazzling showpiece unmatched by anything else visible in the telescope. No one will forget that first telescopic view of Saturn — the chilling beauty of the small, pale orb and the delicate encircling rings magically floating in a field of black velvet. Even the smallest refractor distinctly reveals the rings, although we are gazing across nearly a billion miles of space. The image of Saturn in a quality 8-inch telescope sometimes exceeds the finest photographic portraits of the planet.

The rings of Saturn are enormous in extent — from one edge to the other, they measure more than two-thirds the distance from Earth to the moon. They con-

King of the planets and ruler of 16 satellites, mighty Jupiter is a gigantic sphere of hydrogen and helium gas topped by colourful belts of ammonia-ice clouds. At least two of the dark cloud belts are visible in any small telescope.

sist of trillions of tiny ice moonlets whirling about Saturn, each in its own orbit, in periods ranging from 4 hours for the inner edge to 14 hours for the outer edge. In the densest sectors, the moonlets — from dust particles to house-sized boulders — form a dazzling celestial blizzard around the planet. Yet if an astronaut were within the rings, there would be little danger as long as he were orbiting Saturn at the same velocity. The debris would then move in stately procession, with few collisions.

From a distance, the rings appear as a solid sheet, precisely aligned by gravitational forces into a zone less than a mile thick, exactly above Saturn's equator. In apparent size, the rings are about as wide as Jupiter's disc. Saturn itself appears as a pale yellow globe about one-third the width of Jupiter. Sometimes, one or two dusky bands are seen on the planet's equatorial region. Findings from the Voyager spacecraft missions to Saturn revealed superhurricane wind belts similar to Jupiter's, but an upper-atmosphere ice-crystal haze results in a subdued surface as viewed from Earth. Saturn rotates on its axis in 10 hours 40 minutes, but the planet's bland face makes the spin imperceptible.

The rings of Saturn are more reflective than the clouds that enshroud the planet and are therefore noticeably brighter. Gaps and brightness differences define several distinct rings, only two of which are visible in backyard telescopes. The pair are separated from each other by Cassini's division, a gap about as wide as the United States. This division can be seen with a 3-inch refractor under good conditions. Often easier to detect is the shadow of Saturn on the rings as they curve behind the planet. Also look for the rings' shadow on the disc of Saturn itself.

Saturn controls a family of at least 17 moons. Titan, the largest, is slightly bigger than Mercury and is cloaked in a nitrogen blanket denser than the Earth's atmosphere. Titan could be regarded as a small planet orbiting a large planet. The big moon is visible in any telescope as an eighth-magnitude object looping around Saturn in 16 days. At maximum distance east or west of Saturn, Titan appears about five ring diameters from the planet. Charting the path of Titan around Saturn is a good project for the beginning astronomy enthusiast. Like all the inner satellites in Saturn's system, Titan orbits exactly above the planet's equator. Of the remaining moons, Rhea, at 10th magnitude, is the only one readily seen in telescopes less than 6 inches in aperture. Its orbit lies well within Titan's, less than two ring diameters from Saturn. Two other moons, Dione and Tethys, are targets for 6-inch and larger telescopes. They are often nestled in the glare just beyond the rings, but finding them is one of the bonuses in Saturn observing. Try using averted vision, as explained in Chapter 6.

From one year to the next, Saturn's rings noticeably vary their inclination, ranging from edge-on in 1980 and 1995 to a maximum tilt of 27 degrees in 1988 and 2008. Saturn is the most oblate planet in the solar system, bulging at the equator by an amount equal to the Earth's diameter. This difference is obvious when the rings are within a year or two of being edge-on.

THE OUTER PLANETS

Beyond Saturn, the solar system becomes bleak. Uranus, Neptune and Pluto are not only remote but also dim due to their residence in the gloom so far from the sun. Uranus is visible to the unaided eye, but barely. At magnitude 5.8, it is at the threshold of vision. However, binoculars, in conjunction with the guide map published in the *Observer's Handbook*, easily show the seventh planet. I have had no trouble spotting it from a city-apartment balcony using 7 x 50s. But seeing something more than a starlike dot is another matter. A good telescope and a power of 100 or more are needed to make the little bluish green disc obviously nonstellar. Even its largest moons (less than half the size of Jupiter's) are invisible in all but the most sophisticated amateur equipment. Unlike Saturn, which gains more than a magnitude in brightness when its rings are tipped toward us, Uranus benefits not a smidgen from its necklace of narrow, black rings. In fact, Uranus's rings are so dim, they have never been visually detected from Earth (only electronically amplified images reveal them).

Uranus is so inconspicuous that it was mistaken for a star dozens of times before its accidental discovery in 1781 by English amateur astronomer William Herschel using a 6-inch Newtonian reflector. Since that time, various observers using large telescopes have reported pale belts parallel to the equator. Theoretically, it is possible to see detail on Uranus; at 450x, the planet appears the same size as the Earth's moon does to the unaided eye. But Voyager 2 photographs taken during the January 1986 flyby of Uranus reveal nothing but a uniform aquamarine smog deck that conceals whatever cloud belts exist deeper in the atmosphere.

Neptune, half Uranus's apparent size, is an even tougher object to discern as nonstellar in a telescope. Finding it is enough reward for most backyard astronomers. At magnitude 7.7, it is also a binocular object, but identification is hampered somewhat by large numbers of seventh- and eighth-magnitude stars in Sagittarius, where the planet is located for the next few years.

Pluto reached its closest possible point to Earth in May 1990 — closer than it has been at any time since 1742, centuries before its discovery in 1930. Therefore, the next few years mark the best possible time to attempt a sighting. Just seeing Pluto is a major achievement that few amateur astronomers can claim. It is so dim, magnitude 13.7, that a 6-inch telescope is minimum equipment. A larger instrument makes the task easier but by no means simple. Use the charts in the *Observer's Handbook* or *Sky & Telescope* magazine. To confirm a sighting, two observations are necessary to reveal Pluto's motion among the stars. Otherwise, Pluto cannot be distinguished from a star.

On June 14, 1991, Venus, Jupiter and Mars (in order of brightness) joined in a memorable cluster that impressed millions of astronomy fans worldwide. **Left:** *Superb drawing of Saturn made by English amateur astronomer Paul Doherty based on his observations with a 16-inch Newtonian reflector.*

8

Moon & Sun

It is the very error of the moon;
She comes more near the Earth than she was wont,
And makes men mad.

— William Shakespeare

Seventeenth-century Italian scientist Galileo was, so far as we know, the first to use a then new invention called the telescope to peek at celestial objects. When he turned his instrument to the moon, he was astounded. "The moon is not smooth and uniform," he wrote, "but is uneven, rough and full of cavities." With one glance, Galileo had smashed the centuries-old belief that heavenly bodies were smooth, uniform and precisely spherical.

Times have changed, but at least part of the rush of amazement that Galileo must have experienced when he initially looked at the moon through a telescope is felt by everyone who sees the lunar surface close up for the first time. Even with binoculars, the view is startlingly sharp and clear, with a few dozen craters and rugged mountain peaks plainly visible.

No matter what instrument is used, lunar detail is most distinct along the terminator, the line dividing the illuminated and the unilluminated portions of the moon. This is because of the sharp relief effect caused by the shadows. Contrary to expectations, the full moon is the worst phase to observe because the strong relief effect is absent. As the moon nears full, light-coloured rayed craters stand out like splashes of white paint, but most other surface features are lost in the wash of light. The days around the first and last quarter are best for detailed viewing.

The moon is so near that any telescope will show a wealth of detail. For example, a 60mm (2.4-inch) refractor at 40x or 50x will detect craters about four miles across. With every increase in telescope size, up to about an 8-inch aperture, more and more detail is revealed, down to a limit of about half a mile, or one kilometre, wide. However, shadows cast by linear features as small as 100 feet high can be glimpsed at the terminator. Larger telescopes seldom reveal more, due to seeing limitations.

Astronomers have known for more than a century that the moon's surface has been an unchanging vista. The craters and plains that we see are the same ones Galileo looked at and cavemen wondered about when pondering the nature of the silvery orb that rides the night sky. But in that constant vista, there is much to see. A good telescope offers a lifetime of lunar exploration. There is a magical sense of discovery in following the crawling terminator hour to hour, night to night, as the phase advances.

While the four main phases of the moon — first quarter, full, last quarter and new — are enough detail for wall calendars, backyard astronomers describe the phases in terms of the moon's "age" in days after new phase (new being the point when the moon is closest to being between Earth and the sun). One complete cycle from new to new takes 29½ days. First quarter occurs at 7 days, full moon at 14 or 15 and last quarter at 22.

There are two basic classifications of features on the lunar surface — the familiar craters and the darker plains. The craters are named after prominent philosophers and scientists of the past. The plains are officially called seas, or *mare* in Latin, because they were assumed to be bodies of water by Galileo and other 17th-

century lunar observers. Mare Crisium, best seen when the moon is three to five days old, is a particularly distinctive plain because of its dark tone and circular appearance. Current theory suggests that it was once a vast crater that flooded with lava early in the moon's history.

Around first quarter phase, age six to nine days, the moon is at its prime for backyard skywatchers. About 35 percent of the surface consists of dark, smooth plains. Many bear high-flown names, inherited from the 17th century. Mare Tranquillitatis, the sea of tranquillity, was the site of the first moon walk ("Tranquillity Base here, the Eagle has landed"). The landing site was not far from the crater Maskelyne. Joined to Tranquillitatis is the smaller Mare Nectaris, whose southern edge inundates the largely ruined crater Fracastorius. On the edge of Nectaris, nearest the moon's centre, are three huge craters that provide a perfect contrast to the plains of Nectaris. This region is breathtaking when near the terminator.

The most impressive of the three craters is 67-mile-wide Theophilus, one of the truly great lunar craters. The walls of Theophilus soar to heights more than 21,000 feet above its floor and cast magnificent shadows that intensify its dark, rugged appearance. When the moon is five to six days old, the terminator plunges the crater deep in shadow. I have caught it with only the towering central peak illuminated, the inky blackness giving an eerie and totally unearthly aura to the great crater. When Theophilus was formed a few billion years ago, it smashed down a wall of its nearest neighbour, the equally wide Cyrillus, whose ramparts are now only half as high. Forming the final member of the trio is Catharina. Catharina is about the same size as Theophilus but is in worse shape, its crumbled walls only 9,000 feet higher than its rubble-strewn floor. These three craters, each in a different stage of decay due to the various forces that have modified the lunar surface throughout its history, provide a vivid cross section of crater forms.

Vast lunar plains like Tranquillitatis

and Serenitatis are believed to be the filled-in remains of enormous craters that were blasted out of the lunar surface by impacting asteroids during the early days of the solar system. Rippling across the floor of these two Ohio-sized plains are frozen ridges resembling waves of lava that cooled and then froze in place. This is almost certainly what they are. The low sun angle makes the floor of Serenitatis look as if it is riddled with the underground trails of some monstrous lunar rodent.

After first quarter moon (9 to 11 days old), the terminator sweeps over some spectacular lunar scenery. This phase, known as gibbous, which means humpbacked (due to the convex curvature of the terminator), reveals the rugged southern section of the visible lunar disc. Although the craters are jammed together in bewildering profusion, the monster crater Clavius, 144 miles across, clearly stands out as the largest. Clavius is so vast that it has two 30-mile-wide craters within it, as well as a number of easily seen smaller ones. But these intruders barely make inroads into this colossal lunar feature. The giant walls of Clavius range up to three miles above its floor. Yet the slope of the walls is gentler than it appears. The walls are 35 miles wide at the base, so the wall slope angle is only 10 degrees. This crater can be spotted with binoculars.

Another prominent crater in this rugged sector, Tycho, measures a little more than 50 miles wide, with ramparts about 2½ miles above the floor. The walls rise more steeply than those of most craters, giving Tycho a jagged appearance when the terminator is nearby. When examining the massive walls and sharp peaks of Tycho and similar craters, one gets the feeling that they must be spectacularly impressive from the surface of the moon. Alien, yes, but compared with the Earth's major mountain ranges, like the Rockies, even the steepest crater walls are subtle hills. There may once have been such peaks on the moon, but billions of years of pounding from impacting debris has hammered them down. In addition, the lunar surface material, which is comparable to lumpy garden dirt, tends to slump over time. This is evident in the terraced walls of craters. Tycho is believed to be the youngest of the large lunar craters, and consequently, its walls are among the steepest. Even so, it is hundreds of millions of years old, far older than all of the steep ranges on Earth, which emerged from the action of plate tectonics (continental drift), a process completely absent on the moon.

As the moon advances to full, Tycho seems to get brighter, while the jumble of craters around it declines. Tycho is at the focus of a rayed spray that is evident in binoculars near full moon. Tycho resembles the pole on a globe, with white lines of longitude radiating from it. This is because the asteroid impact that created Tycho splashed debris in all directions – in some cases, a third of the way around the moon. The rays radiating from Tycho are lighter than the lunar surface they cover because solar radiation darkens moon dirt as time passes. Tycho is relatively young, perhaps half a billion years old, less than one-sixth as old as almost every other feature seen on the moon. However, the rayed phenomenon shows up only as the moon approaches full. When the terminator is near Tycho, little of the prominent ray system is seen.

Near the terminator, when the moon is 9 to 10 days old, lies the splendid Copernicus crater, considered by many observers to be the most awesome lunar feature. At only 60 miles wide, it is certainly not the largest crater on the moon. What gives Copernicus prominence is its position in an otherwise flat lunar plain. The contrast is striking. The great two-mile-high walls of Copernicus and a blanket of debris splashed out around it make this the prototype of craters. Imagine the processes that occurred when a multi-billion-ton asteroid slammed into the moon, gouging out this feature perhaps 700 million years ago. When Copernicus is close to the terminator, ragged, inky shadows from its peaks drape over the crater's interior and spill onto the surrounding plain.

Billions of years ago, the moon was much closer to Earth than it is today. Consequently, our planet's gravitational pull on the moon was stronger, raising substantial tides on the lunar surface. This force gradually retarded the moon's spin on its axis, eventually bringing it to a halt with its slightly more massive side locked toward us. No matter what the moon's phase or what season of the year, we never view the other side. Speculation about what secrets the hidden side of the moon contained ended when the first spacecraft looped around our satellite in 1959 and found a cratered landscape similar to the side we see from Earth.

The hidden far side of the moon is frequently — and incorrectly — called the dark side. The dark side is merely the portion of the moon experiencing nighttime. The far side and the dark side coincide at full moon. At other lunar phases, portions of the near side are also dark. The complete cycle of phases repeats every 29.53 days, with all places on the moon in darkness for half that interval.

Probably the best-known but least-understood lunar phenomenon is the harvest moon, the full moon that occurs nearest the autumnal equinox, the first day of autumn in the northern hemisphere, usually September 22. For observers in the northern hemisphere, the full moon nearest the autumnal equinox seems to linger in the sky night after night. To farmers of earlier generations, the harvest moon was an unexplained but welcome bonus of light.

From night to night, the moon moves about 12 degrees eastward, which means that the moon rises in the east an average of 50 minutes later each day. However, the geometry of the orbit of the moon related to the tilt of the Earth's axis results in the moon moving in its orbit along a trajectory nearly parallel to our horizon on the days near the equinox. This means that the time interval between successive risings is much shorter than average.

In southern Canada and the northern United States, the moon rises about 25 minutes later each night around the equinox. Thus for several nights near full

The full moon, **above,** *displays the spectacular rayed crater Tycho in the central lower sector.* **Left:** *The crescent phase, two or three days after new moon, is the time to see earthshine, sometimes called "the old moon in the new moon's arms." Sunlight reflecting off Earth weakly illuminates the moon's night side. The phenomenon is plainly visible to the unaided eye, but binoculars offer the best views. This eight-second photograph overexposed the crescent to show lunar night-side detail.*

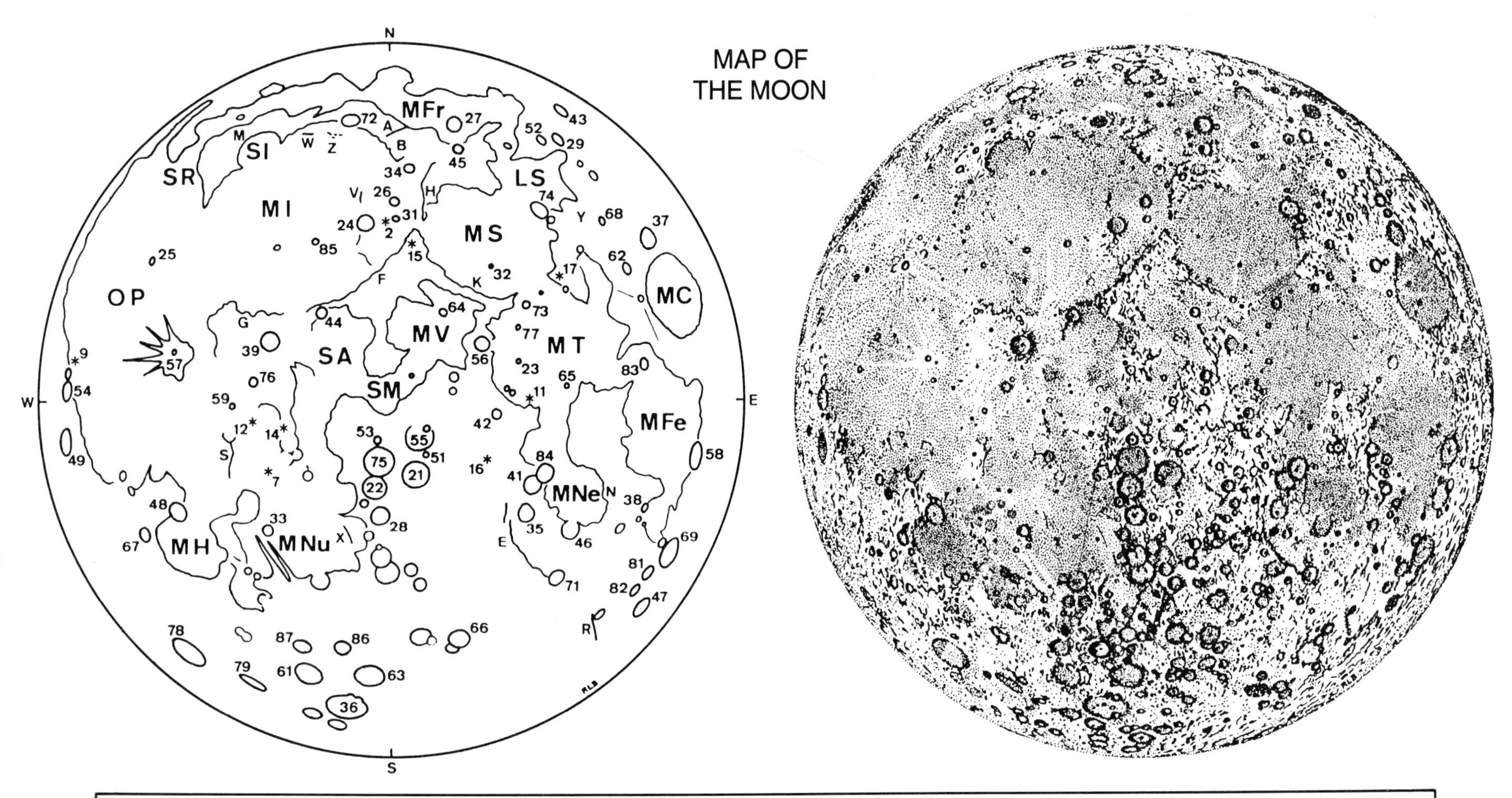

Key to the Map of the Moon

Craters

21 – Albategnius
22 – Alphonsus
23 – Arago
24 – Archimedes
25 – Aristarchus
26 – Aristillus
27 – Aristoteles
28 – Arzachel
29 – Atlas
31 – Autolycus
32 – Bessel
33 – Bullialdus
34 – Cassini
35 – Catharina
36 – Clavius
37 – Cleomedes
38 – Cook
39 – Copernicus
41 – Cyrillus
42 – Delambre
43 – Endymion
44 – Eratosthenes
45 – Eudoxus
46 – Fracastorius
47 – Furnerius
48 – Gassendi
49 – Grimaldi
51 – Halley
52 – Hercules
53 – Herschel
54 – Hevelius
55 – Hipparchus
56 – Julius Caesar
57 – Kepler
58 – Langrenus
59 – Lansberg
61 – Longomontanus
62 – Macrobius
63 – Maginus
64 – Manilius
65 – Maskelyne
66 – Maurolycus
67 – Mersenius
68 – Newcomb
69 – Petavius
71 – Piccolomini
72 – Plato
73 – Plinius
74 – Posidonius
75 – Ptolemaeus
76 – Reinhold
77 – Ross
78 – Schickard
79 – Schiller
81 – Snellius
82 – Stevinus
83 – Taruntius
84 – Theophilus
85 – Timocharis
86 – Tycho
87 – Wilhelm

Mountains

A – Alpine Valley
B – Alps Mts.
E – Altai Mts.
F – Apennine Mts.
G – Carpathian Mts.
H – Caucasus Mts.
K – Haemus Mts.
M – Jura Mts.
N – Pyrenees Mts.
R – Rheita Valley
S – Riphaeus Mts.
V – Spitzbergen
W – Straight Range
X – Straight Wall
Y – Taurus Mts.
Z – Teneriffe Mts.

Maria

LS – Lacus Somniorum (Lake of Dreams)
MC – Mare Crisium (Sea of Crises)
MFe – Mare Fecunditatis (Sea of Fertility)
MFr – Mare Frigoris (Sea of Cold)
MH – Mare Humorum (Sea of Moisture)
MI – Mare Imbrium (Sea of Rains)
MNe – Mare Nectaris (Sea of Nectar)
MNu – Mare Nubium (Sea of Clouds)
MS – Mare Serenitatis (Sea of Serenity)
MT – Mare Tranquillitatis (Sea of Tranquillity)
MV – Mare Vaporum (Sea of Vapours)
OP – Oceanus Procellarum (Ocean of Storms)
SA – Sinus Aestuum (Seething Bay)
SI – Sinus Iridum (Bay of Rainbows)
SM – Sinus Medii (Central Bay)
SR – Sinus Roris (Bay of Dew)

Lunar Probes

2 – Luna 2, first to reach moon (9/13/59)
7 – Ranger 7, first close pictures (7/31/64)
9 – Luna 9, first soft landing (2/3/66)
11 – Apollo 11, first men on moon (7/20/69)
12 – Apollo 12 (11/19/69)
14 – Apollo 14 (2/5/71)
15 – Apollo 15 (7/30/71)
16 – Apollo 16 (4/21/72)
17 – Apollo 17 (12/11/72)

moon, there will be bright moonlight in the early evening when harvesters can make the best use of it.

Sightseeing on the moon from your own backyard is one of amateur astronomy's most accessible pastimes. The moon can be observed just as well from city or country. The late Isaac Asimov, a science and science fiction author who lived in a 33rd-floor apartment in Manhattan, viewed the moon perfectly well with a telescope on his balcony. One early morning when he passed a window on his way to the kitchen, he had this reaction: "Looking out the westward window, I saw it: a fat, yellow disc in an even slate-blue background, hanging motionless over the city. I found myself marvelling at Earth's good fortune in having a moon so large and so beautiful."

OBSERVING THE SUN

What would the average star look like close up? Probably not much different from the sun, whose surface can be examined by any backyard astronomer. But this should not be attempted without taking strict precautions. Concentrated sunlight streaming through a telescope's ocular can cause blindness in less than a second.

There are two chief methods of solar observation that are perfectly safe. The first procedure requires a full-aperture solar filter, which intercepts the sun's light before it enters the telescope, reducing the sun's brightness by a factor of about 100,000. This brings the light intensity down to a safe, comfortable level. Such filters are attached to the telescope in front of the objective lens or mirror.

Full-aperture filters are made of coated optical glass or highly reflective coated Mylar film. The glass filters, though more expensive, yield close-to-true-colour images of the sun. The less costly Mylar filters usually give the sun a bluish tint, although there are some close-to-true-colour transmission Mylar filters available. Full-aperture filters are usually not supplied as standard equipment with telescopes. However, some telescopes come

Small refractors are ideal for projecting a bright image of the sun. In this instance, a partial solar eclipse was under way.

equipped with a small filter designed to be placed at either end of the eyepiece — capped on the end closest to the eye or screwed or clipped into the end that fits into the focusing sleeve.

I do not recommend the use of these eyepiece filters. They quickly heat up from concentrated solar radiation near the focus of the telescope and, after a few minutes, crack or melt, allowing intense radiation to strike the observer's eye suddenly. This happened to me. The filter supplied with my first telescope (the cheap one) fractured after about 10 minutes of use. Fortunately, I instinctively turned away before any eye damage was done. It surprises me that these dangerous accessories have not been banned by consumer-protection legislation.

The alternative to using filters is the indirect viewing method, whereby the solar image is projected by the telescope onto a white viewing card. Binoculars stabilized on a tripod may also be used. This technique has an advantage over filtration because it does not require any accessory equipment. The telescope system is used as is. The intensity of sunlight is more than sufficient to produce a readily visible image that can be sharply focused with adjustments on the telescope. Keep an eye on any children who are enjoying the activity. They are at the right height and have enough curiosity to try to look through the telescope. It is also a good idea to cover the objective of the finderscope, since it projects its own image. Projection usually offers less detail for solar-surface observation than the filtration method, but it has the advantage of providing an image for several onlookers. Hold the projection card one to two feet behind the eyepiece to get the best combination of image size (a 3-inch diameter is good) and brightness.

Small refractor telescopes are best for projection. Bigger instruments are badly affected by seeing caused by solar heating of the atmosphere. Larger telescopes operate better for solar work if a piece of cardboard with a 2½-inch hole is taped in front of the instrument. This reduces the seeing effects and prevents the optical system from overheating, which can be especially dangerous with Schmidt-Cassegrains. Intense heat can also damage expensive eyepieces, another good reason to reduce the aperture of any telescope to about 2½ inches.

The projection card can be hand-held or attached to a rod or stick that, in turn, is fastened to the telescope. This frees the observer from holding the card. To increase image contrast, another card should be attached to the telescope tube so that it casts a shadow on the card receiving the sun's image. Find the sun by adjusting the telescope tube until it makes the smallest possible shadow on the ground. The sun should then be nearly centred in the telescope, and probably only a slight adjustment will then be needed to bring the image onto the screen. Do not sight along the telescope tube.

For unaided-eye observation, a conveniently available and perfectly safe filter is the glass insert from welders' goggles.

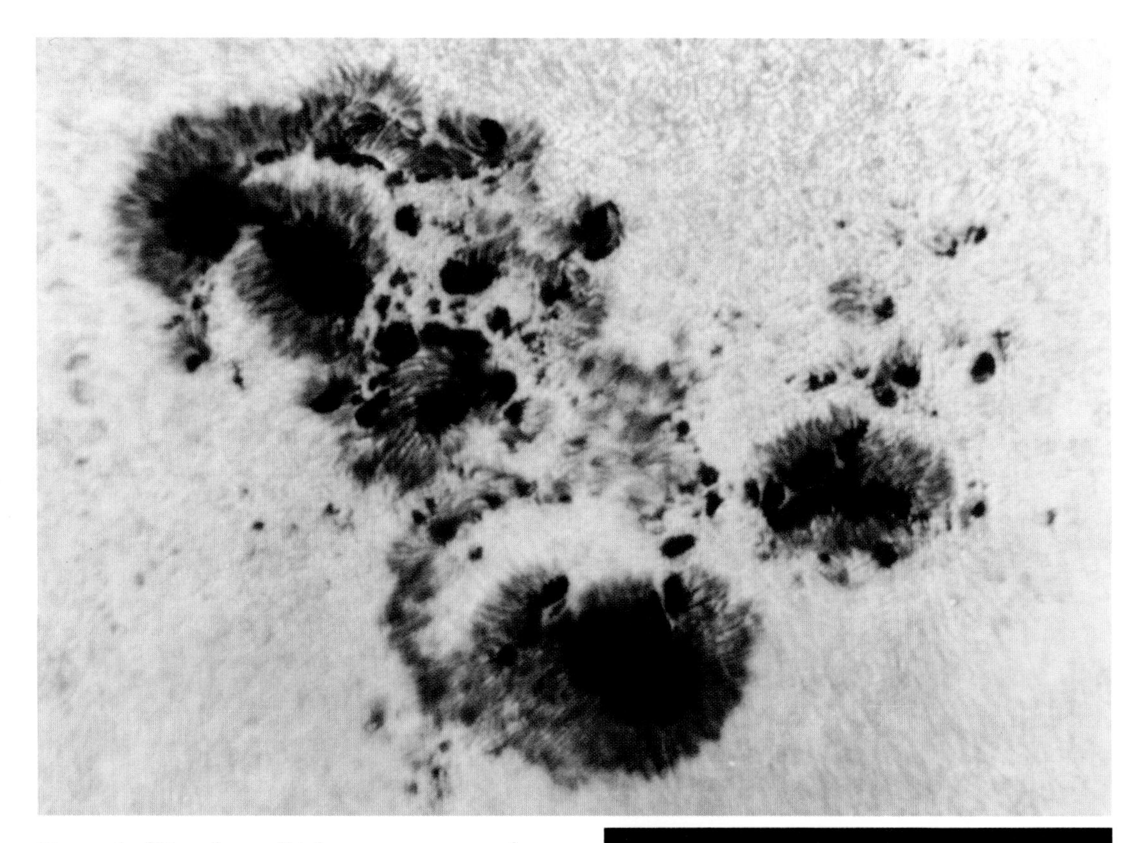

Properly filtered, small telescopes can reveal an amazing amount of detail on the sun, our nearest star. Sunspots are the most obvious solar activity, seen close up, **above**, *in a photograph taken in July 1988 with a 7-inch refractor. Lower of the two full-disc photographs,* **right**, *was taken the same day. Upper photograph, obtained a week earlier, shows the effect of the sun's rotation. Both images were taken with the 4-inch refractor,* **far right**, *equipped with a nickel-chromium-coated glass filter made for visual observation. Special narrow-band filters mentioned in the text produce visual images of striking detail,* **far right**, *showing prominences on the limb and the sun's churning surface. Photograph is a composite of two 8-inch Schmidt-Cassegrain images combined to simulate filter's visual performance.*

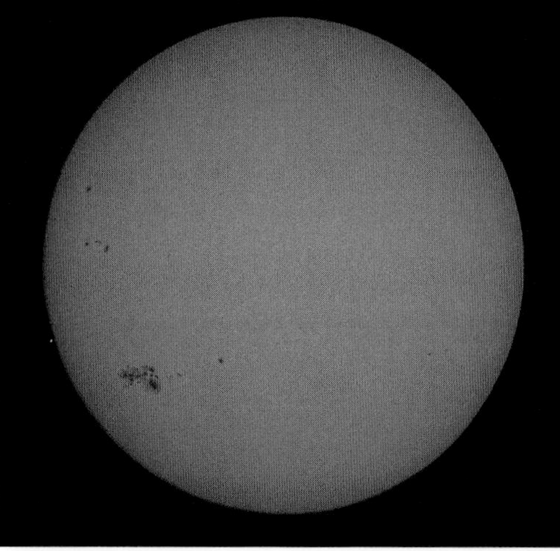

Ask for No. 14 welders' plate at a local welders' supply outlet. It comes in a handy 2-by-4-inch rectangle that permits direct viewing of the sun with both eyes. This filter material can be taped over the objective lenses of binoculars with 40mm or smaller objectives. However, it is *not* safe to use at the eyepiece end, where the light is concentrated. To be effective, *all* filters must diminish the solar intensity *before* it enters the optical system.

With the unaided eye protected by a welders' filter, tiny black spots can often be picked out on the dazzling solar disc. These are sunspots. The fact that they can be seen with the unaided eye and a proper filter is one of the best-kept secrets of backyard astronomy. But anyone with average vision should be able to see a major spot if it is there. I have seen spots in this way many times.

Sunspots have been observed with the unaided eye for at least a thousand years. Chinese astronomers noted black speckles on the solar disc near sunset, when its image is reddened and subdued by atmospheric absorption. Do not try to repeat the Chinese observations if you have to squint. The sun must be *deep* orange.

Binoculars (shielded) almost always reveal a spot or two, and a small telescope will show all the spots on the solar face — perhaps a dozen or more. A magnification of 40 or less is all that is needed for a good view of the entire sun. Higher powers show less than the full solar disc, and when the projection method is used, they render the sun's image too dim to be viewed clearly on the projection screen. It is amazing what can be seen with low magnification. Since the sun is a colossal globe of gases (mostly hydrogen), the visible outer surface is constantly changing. (The sun's diameter is 109 times the Earth's.)

A sharply focused telescopic image of the solar disc will reveal several sunspots, some bright patches and a gradual darkening toward the edge, called limb darkening. With good seeing conditions, the sun's surface appears mottled or granulated, like leather when closely examined. This solar granulation is real and consists of cells of rising gas boiling like water in a pot. A solar granule is about the size of Lake Superior, and it changes its shape in a matter of minutes. The specific change is not evident because it is impossible to concentrate on a single granule for more than a few seconds. The pattern overwhelms each individual spot. The darkened limb of the sun is usually flecked by irregular, bright splotches called faculae. These are clouds of hydrogen, often associated with sunspots, surging above the sun's surface. Lack of contrast prevents them from being visible near the centre of the disc.

The major fascination of the sun lies in those black blemishes, the sunspots. The spots themselves are actually cooler areas on our star, about 1,500 degrees C cooler than the 5,500-degree solar surface. Being cooler, the spots are darker, but they appear black only by contrast. They are actually a light brown colour and would appear so if seen alone.

There are two parts to a sunspot: a black, virtually featureless interior, called the umbra, and a greyish, feathery structured zone around the umbra, called the penumbra. Combined, these two regions come under the general designation of a sunspot. Sunspots can be 10 or more times the size of Earth. An average one is at least as big as our planet.

Like all visible solar features, the spots are caused by intense magnetic fields that coil within the sun and break out through the surface in largely unpredictable ways. A sunspot is the focus of a magnetic-field breakout, a region where the flow of energy from within the sun is restricted, hence the less luminous appearance. An individual spot emerges as if from nowhere over a period of a day or two. Spots can last for days or weeks. Some of the largest remain visible for several months.

The number of spots varies from year to year in a fairly well-defined 11-year period, known as the sunspot cycle. These cycles do not exactly repeat one another; some are more "spotty" than others. In general, the rise to maximum is more rapid (about four years) than the drop to minimum (about seven years). The most recent minimum occurred in 1986. The mechanism behind the sunspot cycle remains a scientific enigma. Spots do not appear over the sun at random. They are usually confined to a region 15 to 40 degrees north and south of the solar equator. At sunspot maximum, the spots are generally closer to the equator than they are earlier in the cycle.

Following the progress of sunspots over a period of weeks or months is easy. Each time the sun is visible, make a sketch of the sunspots' positions. Place the observing sheet on a clipboard, and hold it behind the telescope. This should result in an accurate enough drawing to show positional changes, as well as the growth and decay of individual spots. From one clear day to the next, these drawings will show the progress of the spots as the sun's rotation carries them around the visible disc. (The sun rotates once in about 27 days.) To ensure that each picture has the same orientation, note the direction of the sun's drift through the field of view caused by the Earth's rotation. The point on the sun's disc that first drifts out of the field of view will establish a reference mark on the circumference of each circle, permitting proper relationship of one drawing to the next.

Solar observing ranks as one of the most pleasant astronomical activities because it is conducted on sunny days and requires only modest equipment. But the joys of solar observing have really hooked some amateurs. Not content with standard methods of watching the sun, they equip their telescopes with narrow-band filters that permit the ultimate — watching solar prominences, those "flames" seen in the most spectacular photographs of the sun. These filters range from several hundred to several thousand dollars. The performance of the more expensive ultranarrow-band filters is awesome. The first time that I saw the sun through one, I could not tear myself away. Our familiar star was transformed into a ball of roiling plumage as intricate in its detail as the rugged face of the moon.

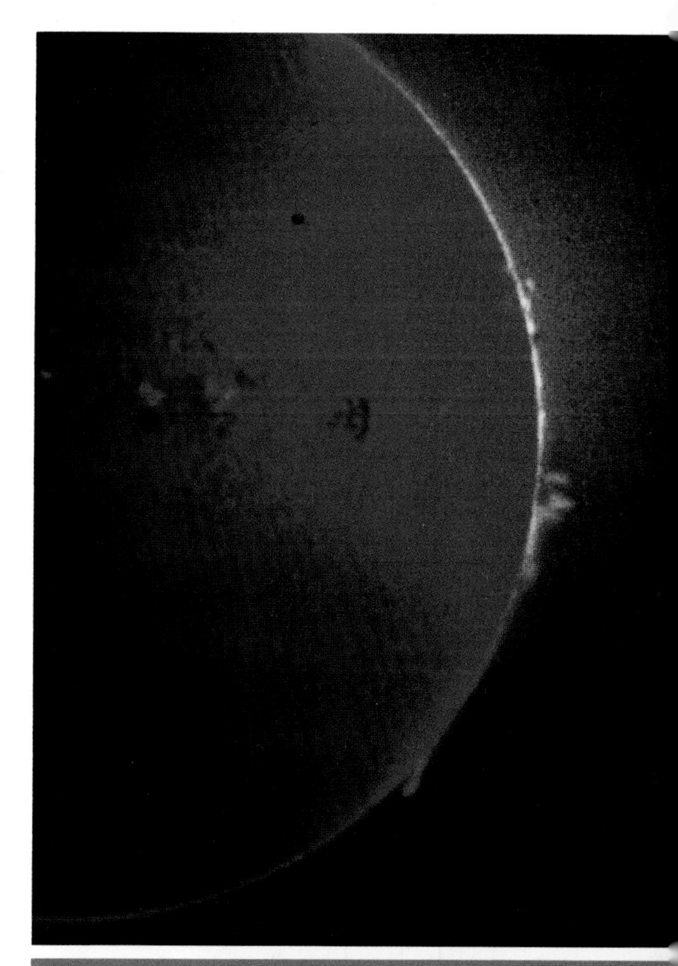

THE HORIZON ILLUSION

Have you ever noticed how the moon appears larger when it is near the horizon than when it is overhead? The difference is so apparent, it seems almost impossible that it is not real. Yet how could it be? The moon is no closer to us on the horizon than it is overhead. Actually, it is about 4,000 miles farther away (as we must look across the radius of Earth).

The same effect occurs with the sun. It seems enormous as it dips below the horizon — a fiery ball reddened by intervening dust and particles in the Earth's atmosphere. At these times, the sun is distorted to an oval by the refractive properties of the atmosphere that bend light rays, similar to the way a straight stick appears bent when partially submerged in water.

Could this refraction be enlarging the sun's apparent size? Simple tests have shown that the reverse occurs. The atmosphere acts like a weak lens, compressing the sun's vertical dimension and giving it an oval shape that is smaller than the circular disc would appear. The same phenomenon can be seen when the full moon is close to the horizon. It appears like a huge cosmic pumpkin, the reddening caused by the same atmospheric dust and haze that redden the sun.

Since neither atmospheric refraction nor changes in distance make the moon or sun bigger on the horizon than when either is higher in the sky, why do they seem so large? The effect is particularly noticeable with the moon because it can be compared with its appearance higher in the sky, whereas the sun is too bright for easy comparison. When people unfamiliar with astronomy are quizzed about the moon's size, virtually all of them insist that our satellite is bigger when close to the horizon. Why?

The horizon illusion was recognized as an enigma as long ago as 350 B.C., when Aristotle incorrectly attributed it to atmospheric "vapours" that distort images close to the horizon. Around the year 1000, Arabian physicist Ibn Alhazan offered the first modern explanation. He suggested that a familiar background, such as distant trees or houses, provides a frame of reference not available when the moon is overhead. Since the moon looks huge by comparison with these familiar objects, the mind insists that it is vast in size.

Alhazan's theory sounds plausible, but it fails to explain why the same effect occurs with a perfectly flat desert or ocean horizon. The illusion works even in a planetarium. The projected image of the moon seems bigger near the horizon than when it is higher up on the planetarium dome, even though the lunar image may then be closer to the observer. Obviously, another mental factor is coming into play in addition to comparisons with objects on or near the distant horizon. It was uncovered in 1959 during a test at the University of Wisconsin.

In this test, a 20-inch disc was suspended 85 feet above the ground and another disc the same size was placed 85 feet away on the horizon. People of all ages were asked to stand at a point equidistant from the two discs. The observers did not know that the discs were identical and therefore should appear exactly the same. Everyone tested thought that the disc on the horizon was the larger of the two. Young children most overestimated the difference, some saying that the disc on the horizon looked three or four times bigger than the one overhead.

Somehow, looking up has something to do with the illusion. As a further test, other researchers put volunteers in a dark room with a disc straight ahead and an identical disc at the same distance overhead. Again, everyone thought that the overhead disc was smaller.

So apparently two factors are involved: (a) association with the distant horizon and (b) looking straight ahead as opposed to looking nearly overhead.

Yet there is more to it than that, but nobody is sure just what it is. Even though I am fully aware of the horizon illusion and its various explanations, I still see the full-blown effect. It is probably the most powerful illusion in nature.

There are ways of diminishing the illusion. When the moon is near the horizon, try looking at it through a tube. Without the horizon reference, it seems smaller. Another method of countering the effect is to lie down and look at the moon near the horizon from a position flat on the ground. The moon does not appear nearly as large as it does from a standing position, particularly if the neck is craned to look at it over the head or down toward the feet. Or try standing, bending over from the waist and looking at the moon between your legs. Again, it appears much smaller.

If all this still sounds unconvincing, here is the final test: An aspirin held at arm's length is only slightly larger than the moon. It will cover it nicely, whether the moon is hovering over the horizon or riding high in the night sky. Try it.

Despite what your brain tells you, the moon and the sun are no larger when they are close to the horizon than when they are high in the sky. It is one of nature's most powerful illusions.

9

Solar and Lunar Eclipses

It is only during an eclipse that the man in the moon has a place in the sun.

— Anonymous

I saw my first total eclipse of the sun in February 1979 as a member of the Royal Astronomical Society of Canada's eclipse charter flight, which flew from Toronto to Gimli, Manitoba, the morning of the eclipse. We landed as planned at the unused air force base two hours before the moon took the first notch out of the sun. Although clouds and snow were predicted, the sky was clear, and the jubilant troop of eclipse buffs unloaded about 75 telescopes and more than 100 cameras and set up on the runway.

As the eclipse progressed, the scene began to resemble the climax of the movie *Close Encounters of the Third Kind*, which also involved a battery of equipment on a runway, with scientists and others awaiting the approach of extraterrestrial spaceships. Like the scientists and technicians in the movie, we were not disappointed. As one eclipse watcher said later, it was as if a god had decided to make his appearance for two minutes, and we knew he was coming.

I was completely unprepared for the overwhelming power of the eclipse. About two minutes before totality, the sun's image was reduced to a thin slice along the rim of the black disc of the moon. In a few seconds, I knew that the sun would be gone and that we would be plunged into darkness. Then, like a vast, diffuse storm cloud, the moon's shadow suddenly appeared in the west, growing larger by the second.

With surprising suddenness, the shadow of the moon swept over us, the last rays of sunlight disappeared, and instantly, the sun was transformed into an awesome celestial blossom — the black disc of the moon surrounded by streamers of the sun's atmosphere, the corona.

Peeking around the black disc, and plainly visible to the unaided eye, were a half-dozen solar prominences, like fingers of frozen fire. These surges of hot hydrogen, flamelike in appearance, constantly lurch from the surface of the sun, propelled by immense magnetic fields. I knew they were there, but I did not expect to see them so plainly. (Neither did anyone else, for it turned out that the prominences have never been more spectacular during an eclipse.)

I was so awestruck by all of this that I became incapable of performing even the most rudimentary tasks. I realized immediately that I would never be able to photograph the event — I wanted every second of viewing time to enjoy the image. I wrenched the camera off my telescope and quickly slipped an eyepiece into the focuser. Then I gazed on what has to be the most stunning astronomical spectacle that I have seen in three decades of skywatching.

Through the telescope, I could see detailed structure within the prominences, including one that was completely detached from the surface, like a suspended ball of fire. Judging from its size, it must have been several times wider than the entire Earth.

The most striking feature was the vast range of delicate hues and intricate detail in the corona, the beautiful halo that surrounds the eclipsed sun. It ranges from

a gorgeous pinkish orange close to the sun to various shades of pale yellow, pink and blue farther from the disc. It has an overall brightness comparable to that of the full moon, an eerie yet mesmerizing cosmic flower. The sun's magnetic field twists the corona into feathering arches and swirls that I was able to follow almost a full sun's diameter from the surface of our star.

The moon's black disc looked like a hole in the sky, with a ghostly aura around it and little tongues of pink flame licking its black circumference. Because of some pale ice-crystal cirrus clouds, the sky did not grow as dark as it has during some previous eclipses, or so I was told by some of my colleagues. Nonetheless, the darkness during totality and the colour accompanying it were like a bizarre twilight, with simultaneous dawn and dusk — light in every direction around the horizon and darkness higher in the sky.

In the last few seconds of totality, a glimmer of sun seared between mountain ranges on the edge of the moon's disc. This is the aptly named diamond-ring effect. The diamond ring lasted several seconds, growing from a starlike spot to a dazzling glow.

Then Baily's beads appeared as more sun leaked through the ridges on the moon's limb. After that, the sun was too bright to look at, and it was back to the welders' filters. But by this time, everyone in our group was shouting, cheering and applauding. I was speechless, beside myself with wonder at such an extraordinary visual symphony.

That 1979 eclipse was the last time that the moon's shadow will touch Canada or the United States for 38 years. On average, any place on Earth is treated to a total solar eclipse only once in 360 years.

The moon takes on a coppery glow from sunlight filtering through our atmosphere and refracting into the Earth's shadow during a total eclipse. **Above**, *the exceptionally dark totally eclipsed moon of December 9, 1992, was reduced to one-millionth the brightness of a normal full moon.* **Previous page**, *total solar eclipse, February 1979.*

The arrangement necessary for a total solar eclipse is an exact alignment of Earth, moon and sun, with the moon casting its shadow on Earth. This is a fairly frequent event that occurs almost every year. However, the problem for those who wish to view a total eclipse of the sun is the small size of the moon's shadow. By the time the moon's shadow reaches Earth, it is approximately 100 miles wide. One must be within this shadow in order to see the "darkness at midday," which lasts less than seven minutes.

Only from within the narrow shadow is the sun entirely blotted out by the moon. The sun is replaced by a black disc, the night side of the moon. The exquisite beauty of a total eclipse of the sun can be seen quite safely with the unaided eye or a telescope. (To make this point perfectly clear, no eye protection is needed during the few minutes of totality, when the disc of the sun is completely obscured. The filter methods described earlier are always necessary during any other aspect of the eclipse.)

The key factor is that you have to be within the moon's shadow. Close doesn't count. If just a little bit of the sun remains peeking around the moon's disc, none of the spectacular corona is seen. Because they are visible from a much wider zone, partial eclipses are more common. But a partial eclipse of the sun in no way compares to a total eclipse. During most partials, the dimming of the sunlight is hardly noticeable. Observed through proper filtration, the sun appears to have a nibble taken out of it, like a bite from a cookie. If an image of the sun is projected telescopically or if appropriate filtration is used, the event is interesting and worth observing. Local newspapers should carry timely information about visible partial eclipses (see table at end of chapter).

Why do eclipses of the sun not occur every month at new moon, when the moon is between Earth and sun? Most of the time, the moon passes just above or below the sun and does not obscure it from any position on Earth. The reason for this is that the moon's orbit is slightly inclined to

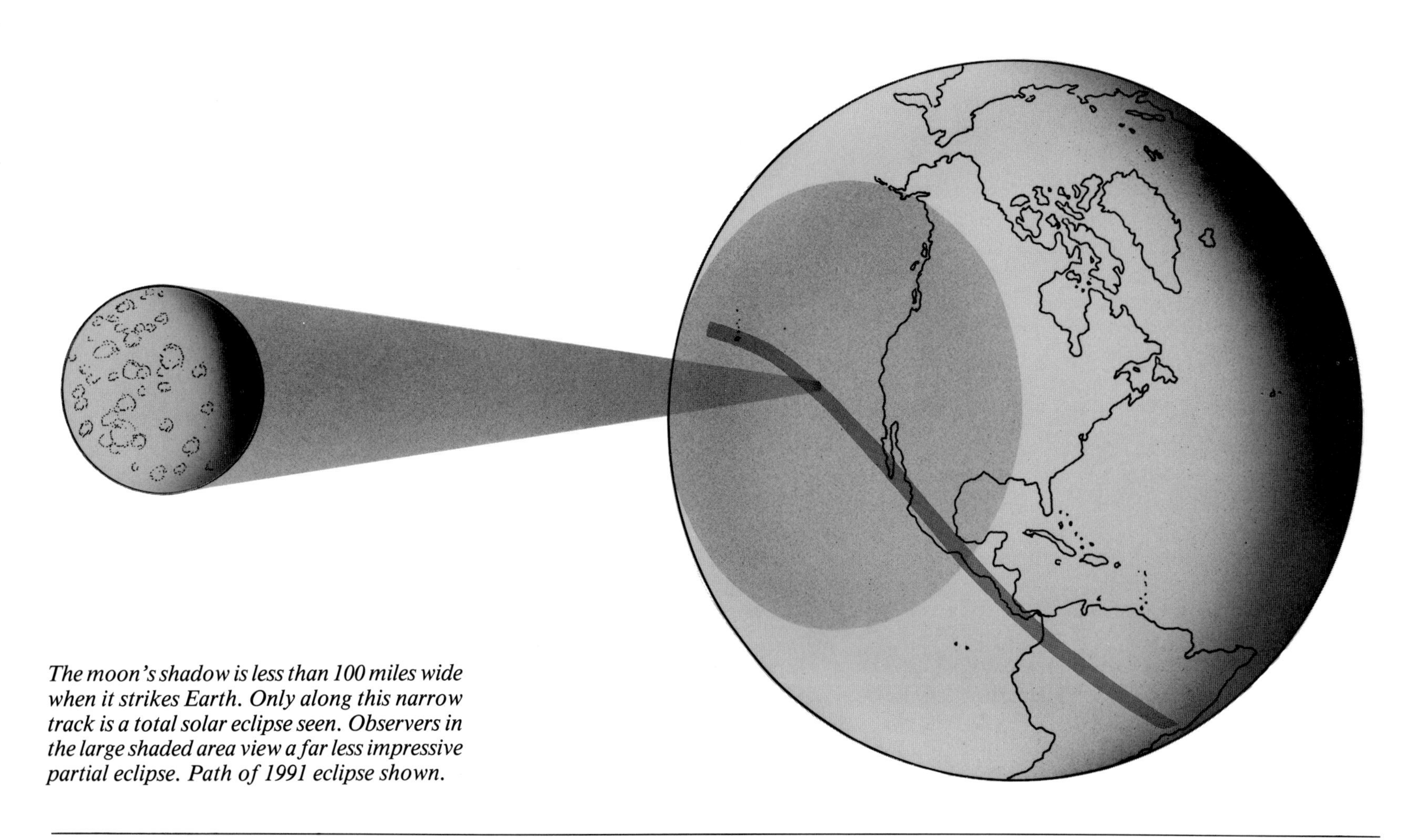

The moon's shadow is less than 100 miles wide when it strikes Earth. Only along this narrow track is a total solar eclipse seen. Observers in the large shaded area view a far less impressive partial eclipse. Path of 1991 eclipse shown.

the ecliptic (by five degrees). This means that except when the apparent path of the sun and the moon intersect, the moon is either below or above the sun, and no eclipse occurs. The same geometry applies to a lunar eclipse.

The earliest recorded eclipse probably occurred on October 22, 2137 B.C. The ancient Chinese chronicle *Shu Ching* tells of two royal astronomers, Hsi and Ho, who were so "drunk in excess of wine" that they failed to warn the populace of the impending darkness. The unexpected eclipse so frightened the people that they stampeded through the streets beating drums to frighten away the dragon devouring the sun.

According to ancient Chinese law, any error by an astronomer in predicting eclipses was of the gravest consequence. If the astronomer's forecasts were "behind the time, he would be hanged without respite." Apparently, that is what happened to Hsi and Ho for drinking on the job. But the tale also suggests that the Chinese could predict eclipses more than 2,000 years before the same method was learned by the Greeks.

It used to be only scientists who journeyed around the globe to gaze at the midday night. The eclipse over Manitoba on June 18, 1860, is an example of the incredible effort that went into such academic jaunts. Simon Newcomb, later to become the most prominent astronomer of his time, headed the 2,000-mile expedition from Boston to The Pas, in northern Manitoba. The trip took five weeks by steamer, covered wagon and canoe. Violent rainstorms and generally bad weather delayed the latter parts of the journey, which could only be made by canoe and portage.

Fearing that they would not make it to the selected site on time, Newcomb persuaded the hired voyageurs to paddle for 36 hours straight to get within the belt of totality. The effort was in vain, for Newcomb and his two assistants simply sat at their telescopes looking at the clouds. As if nature conspired to add a final insult to the frustrated group, the clouds parted just a few minutes after the eclipse had ended.

At one time, eclipses were of enormous scientific importance. Observations that could not be made at any other time were possible during the reduction in solar illumination. Einstein's general relativity theory, which predicted that light would be deflected in a strong gravitational field, was confirmed by observing the position of stars near the eclipsed sun. Stars normally invisible due to the sun's brightness were briefly seen during totality. Photographs of their positions revealed that their light had shifted precisely by the amount predicted in Einstein's theory.

So far as we know, the total-solar-eclipse phenomenon is unique to Earth. It happens because the sun is about 400 times the diameter of the moon but 400 times farther away. Thus they both appear almost exactly the same size. Nowhere else in the solar system does this specific arrangement occur. The satellites of Mars, as seen from the surface of that planet, are too small to cover the solar disc. None of the satellites of Jupiter and Saturn are the same apparent size as the sun, being either significantly larger or smaller. At any rate, those planets are 5 and 10 times farther from the sun, respectively, and any eclipse effect would be much diminished.

Coincidence is truly the only explanation, because the moon has been gradually increasing its distance from Earth over the last three or four billion years. In an estimated 50 million years, total eclipses of the sun will no longer occur because the moon will be too far away. All solar eclipses will then be annular, that is, a small amount of sun will peek around the edge of the lunar disc, ruining the awesome appearance of the event now witnessed. Today, Earth and moon are at the perfect balance, and we are enjoying the maximum effect of an incredible cosmic coincidence.

LUNAR ECLIPSES

In 1503, Christopher Columbus and his crew found themselves stranded on the island of Jamaica, their ships damaged beyond repair. The native Arawak Indians had decided that they were not going to supply any more food in exchange for baubles and trinkets. While the weeks dragged on, morale plunged as the crew was forced to live off the land.

Scanning his navigational tables, Columbus noticed that an eclipse of the moon would occur on February 29, 1504. He craftily devised a plan.

The night of the eclipse, he announced to the Arawaks that the Almighty was frowning on the treatment they had been giving him and his crew. Gesturing to the sky, he reported that God had decided to remove the moon as a sign of his displeasure. Within minutes, the Earth's shadow began to steal across the moon's face.

According to Columbus's diary, the theatrics worked perfectly. The Indians promised to provide all the food the sailors would need if God would only bring back the moon. These histrionics probably prevented Columbus and his men from starving before they were finally rescued and returned to Europe.

An eclipse of the moon can occur only at full moon and only when Earth is exactly between the sun and the moon. When these conditions are met — usually twice a year — the moon is engulfed in the Earth's shadow for up to 1¾ hours. Unlike a total solar eclipse, which is visible only from a restricted zone of totality, a lunar eclipse is seen from the entire night side of Earth — there are millions of good seats available for interested observers.

Our satellite is never completely blacked out during a lunar eclipse. Sunlight diffusing through the Earth's atmosphere bathes the moon in a dull glow that reduces it to about one ten-thousandth of normal full-moon brightness. The same principle causes the early-evening sky to remain relatively bright, even though the sun is below the horizon.

The darkness of the shadow varies with the amount of cloud and atmospheric dust and pollution that surrounds Earth at the time of the eclipse. Occasionally, the clouds are so thick that the shadow is almost black, while at other times, it is a light rust colour. Eruptions from the Mexican volcano El Chichon in April 1982 spewed vast quantities of dust into the upper atmosphere. The dust particles absorbed sunlight, making the two total eclipses later that year among the darkest of the century.

The moon moves eastward (in its orbit around Earth) a distance equal to its own diameter in about an hour. Earth's shadow is just under two moon diameters wide, so if the moon passes through centrally, it can be totally within the shadow for nearly two hours. Sometimes, our satellite just skims the shadow, and a partial eclipse results. The contrast between the bright moon and the dark shadow during a partial eclipse makes the actual tone of the shadow difficult to determine, and the event is far less impressive than a total eclipse.

A total lunar eclipse unfolds as follows: About 20 minutes before the moon is scheduled to enter the shadow zone, the eastern edge of the moon becomes slightly dusky, indicating that the shadow region is nearby. However, when the edge of the moon actually contacts the shadow, the darkening effect is unmistakable. During a total eclipse, the moon takes about an hour to slip into the shadow. Once it is fully immersed, the total eclipse begins. The moon swings out of the shadow during the final hour of the event. The eclipse can last nearly four hours.

At the peak of a total lunar eclipse, the moon ranges anywhere from one percent to one one-hundredth of one percent of its normal brightness — from a light rust colour to almost complete invisibility. The "black" eclipse of 1963 was so dark that some people were unable to find the moon unless shown by someone who knew where to look.

During the 10 minutes centred on mid-totality, a visual estimate of the darkness of the eclipse can be made using a scale developed by French astronomer André-Louis Danjon. This scale is intended for unaided-eye estimates:

Zero on Danjon's scale is a very dark eclipse; moon practically invisible, especially during midtotality. *One* is a dark grey or brownish eclipse; lunar details distinguishable only with difficulty. *Two* is a deep red or rust-coloured eclipse, with the central part of the shadow very dark and the outer edge relatively bright. *Three* is a brick-red eclipse, usually with a bright grey or yellow rim to the Earth's shadow. *Four* is a bright copper-red or orange eclipse, with a bluish, very bright edge to the Earth's shadow.

If the eclipse seems to be between two categories — one and two, for example — it can be recorded as "1.5 on Danjon's scale." Remember, the scale is for unaided-eye estimates; binoculars or telescopes make the eclipsed moon appear brighter.

Large telescopes or high magnifications are of little value for observing a lunar eclipse. I recommend using binoculars or small low-power telescopes, since they allow the entire moon to be viewed during the event.

THE ECLIPSE CULT

Their numbers ebb and swell from one solar eclipse to the next, but they are always there, enduring any hardship or expense to spend a few hundred seconds standing under the moon's shadow. They are the eclipse chasers. No place is too remote: southern Australia in 1976, India in 1980 and Java in 1983. With almost religious fervour, they converge on whatever location the celestial geometry dictates in order to witness nature's superspectacular.

Sometimes, their efforts go unrewarded. Whenever plans for an eclipse expedition are being made, amateur astronomers are reminded of the late J.W. Campbell, an astronomer at the University of Alberta. He travelled around the world to stand under 12 total eclipses during the first half of this century but was clouded out every time.

The success rate in recent years has been much better, because modern weather records allow more judicious site selections and because today's transportation methods offer more flexibility for the expedition. But modern eclipse chasers have their own misadventures.

During his globe-trotting to see nine total eclipses (only one was clouded out), Vancouver astronomy enthusiast Ian McLennan has had his share of unusual mishaps. At a remote site in Kenya the night before an eclipse, a ferocious windstorm whipped the area where his group had camped. After nightfall, McLennan stepped outside his tent for a moment. By the time he returned, the tent had blown away. "I never did find it," he recalls. But he did see the eclipse.

Less fortunate was a planetarium director who, like thousands of others in June 1973, had booked a suite on one of several cruise ships specifically chartered to be under the moon's shadow in the Atlantic Ocean at the critical time. Our hapless eclipse buff had to go below to use the men's room a few minutes before totality. But somehow, he got swallowed up in the labyrinth of hallways in the bowels of the ship. By the time he was able to find his way back on deck, the total eclipse was history.

McLennan's eclipse travels took him to eastern Quebec in 1972, where he had a revealing insight into human nature. People had lugged equipment from all over the globe to observe the eclipse, but when totality arrived, the clouds were too thick to see anything. "We did get plunged into blackness, of course," McLennan recalls, "and we noticed the weird calm that always accompanies totality. Yet on the Trans-Canada Highway nearby, the truckers simply turned on their headlights and kept going! Very few of them stopped to see what was causing the darkness at midday. And we had come from all over the world to experience it."

Eclipse addiction is not limited to amateur astronomers; it afflicts a wide cross section of nature lovers. As one veteran eclipse chaser put it: "It is highly contagious and requires only a single exposure to an unobstructed total eclipse."

Future Solar Eclipses

June 30, 1992:	An ocean eclipse of long duration (5 minutes), spanning the mid-Atlantic.
May 10, 1994:	Although this is only a partial eclipse, it is one that almost everyone reading this will have a chance to see because it sweeps right across North America. Along a 100-mile-wide path running from El Paso, Texas, through Detroit, Buffalo and Halifax, Nova Scotia, spectators will witness a 94% annular eclipse similar to the illustration on the facing page.
November 3, 1994:	The eclipse path cuts through the middle of South America and tracks completely across the South Atlantic but misses South Africa. Totality: 4½ minutes. Seen as a partial in parts of North America.
October 24, 1995:	A 2-minute totality will cross India, Indochina and Indonesia, missing all of the major islands.
March 9, 1997:	Three-minute eclipse over eastern Siberia.
February 26, 1998:	Path crosses southern Panama, Colombia, Venezuela and some Caribbean islands; 4-minute totality.
August 11, 1999:	Millions will see this one, as the path of totality runs through England and Europe. Two minutes.

Partial solar eclipses will be seen in all or part of North America on May 21, 1993; May 10, 1994; October 12, 1996; February 26, 1998; and December 25, 2000.

Future Lunar Eclipses (Visible From North America)

June 15, 1992:	Partial eclipse (69%) visible over the entire continent.
December 9, 1992:	Total eclipse well seen from eastern North America.
November 28, 1993:	Total eclipse visible throughout North America.
May 24, 1994:	Partial eclipse.
April 15, 1995:	Early-morning partial eclipse for western North America only.
April 3, 1996:	Total eclipse visible from eastern North America.
September 26, 1996:	Fine total eclipse for most of Canada and the United States.
March 24, 1997:	Partial eclipse.
July 28, 1999:	Early-morning partial eclipse from western Canada and Alaska only.
January 21, 2000:	Good total eclipse for most of the continent.

The moon can pass precisely between Earth and sun and not produce a total eclipse. These annular eclipses occur when the moon is near its maximum distance from Earth.

10

Comets, Meteors and Auroras

I have watched a dozen comets, hitherto unknown, slowly creep across the sky as each one signed its sweeping flourish in the guest book of the sun.

— Leslie C. Peltier

The evening of April 26, 1978, was not exceptionally good for telescopic observation from the Ottawa, Ontario, area. The sky was clear, but Rolf Meier, an experienced amateur astronomer and a member of the Royal Astronomical Society of Canada, had seen better conditions. However, as he had done on hundreds of previous occasions, he quickly prepared the Society's Ottawa Centre 16-inch telescope for a night's observing. Before moving on to his regular studies of galaxies, he decided to spend a few minutes searching for a new comet. Fifteen minutes later, he found one.

When the discovery was confirmed that night at Lowell Observatory, in Flagstaff, Arizona, Meier joined a select group of successful comet hunters — a rare breed of amateur astronomers who gain cosmic immortality by discovering a comet and having it named after them. Meier had caught his first comet after only 50 hours of searching, a remarkable achievement considering the elusive nature of these celestial objects that Mark Twain called "unaccountable freaks."

Although comets have been observed with the naked eye since antiquity, they still stir excitement when they grace the Earth's sky. Few people on the planet were unaware of the approach of Halley's Comet in 1985. Spacecraft that visited the comet in March 1986 confirmed suspicions that comets are primordial clouds of frozen gases laced with dust and rubble left over from the solar system's birth. These ancient celestial icebergs are almost always far too distant to be seen. But on the rare occasions when they near the sun, they are transformed by solar radiation into ghostly objects, sporting gauzy tails more tenuous than mist but spectacular against the black backdrop of space. The tails are vapours melted from the icy comet nucleus by solar heat.

Most comets appear unexpectedly, as if materializing from nowhere to provide game for comet sleuths like Meier. A typical comet has a huge, elongated orbit, its far end out in the abyss beyond Pluto. At the other end of its orbit, a comet makes a hairpin turn around the sun, emerging with a longer and brighter tail due to the maximum dose of solar heating. It decorates the sky for a few weeks and then retreats into the frigid darkness from whence it came.

According to astronomical tradition, comets are named after their discoverers. This custom has spawned the comet hunters — dedicated amateurs who scan the skies winter and summer just after dusk or before dawn, the times when new comets are most frequently seen. As they gaze through their telescopes, all comet hunters hope to be the first to see one of these solar system renegades.

Comet hunters like Meier know the sky like road maps of their own hometowns. They have learned how to distinguish the faint glow of multi-billion-star galaxies that masquerade as their celestial quarry: a cosmic iceberg just millions of miles, not light-years, distant. Large sectors of the

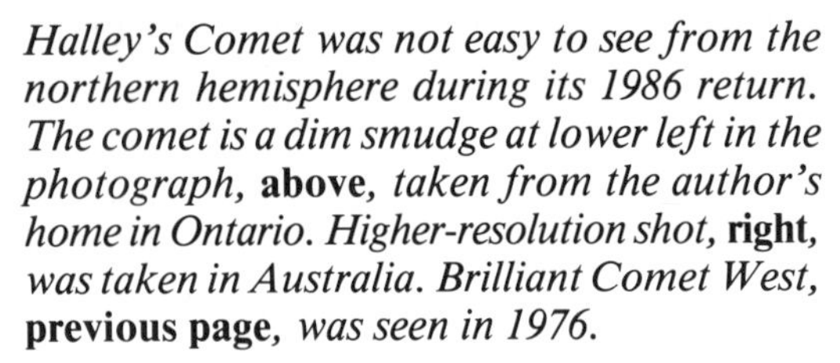
Halley's Comet was not easy to see from the northern hemisphere during its 1986 return. The comet is a dim smudge at lower left in the photograph, **above**, *taken from the author's home in Ontario. Higher-resolution shot,* **right**, *was taken in Australia. Brilliant Comet West,* **previous page**, *was seen in 1976.*

sky are scanned almost every clear night, but comets often elude the eye of the searcher. Many are picked off by professional astronomers engaged in sky photography for other reasons. A major-observatory telescope designed for photography can record a comet long before it becomes bright enough for an amateur's telescope — if it is pointed at the right part of the heavens.

Usually, fewer than a half-dozen comet hunters are rewarded each year, and most of those comets are dim puffs. Only three or four times per century does a really spectacular comet enter our solar system. These marvels become so bright that they can be seen in broad daylight. This is the bounty that all comet hunters seek, but most comets never reach the threshold of naked-eye visibility.

To keep the annual tally of two dozen or so comets straight, the International Astronomical Union, headquartered at Harvard University, has established a system for naming them. The famous Comet Kohoutek, for example, was officially named Comet 1973f ("f" being the sixth comet discovered in 1973). As it turns out, the preceding comet, Comet 1973e, was also discovered by Czechoslovakian astronomer Lubos Kohoutek. To avoid such confusing situations, astronomers adopted the letter system of identification.

Comets are seldom visible for more than a few months, centred on perihelion — their closest approach to the sun. The time of perihelion passage can be determined precisely, and this method is used to provide a comet with its final designation for the record books. Comet Kohoutek 1973f is now known as Comet 1973XII, the twelfth comet to pass perihelion in 1973.

The champion comet discoverer of all time was Jean Louis Pons, a janitor at the Marseilles Observatory, who bagged 37 of the elusive interlopers between 1780 and 1830. He used a 2-inch refractor of mediocre quality for many of his finds — an incredible achievement that could not be duplicated today due to the competition from other comet hunters and from the researchers who inadvertently capture comets on film.

On the average, an easily visible naked-eye comet appears every two or three years. About once a decade, the skies are visited by a naked-eye comet with a bright tail (Comet Bennett 1970, Comet West 1976). Normally, comets are visible only by telescope and are unimpressive sights, resembling faint stars embedded in mist, their tails nothing more than a weak fan of haze. The filmy appendages always point away from the sun, since it is the pressure of sunlight that sweeps the tail away from the nucleus.

The most spectacular comets are the sungrazers, those that come close enough to the solar fires to have their surfaces boiled off in churning fury, thrusting vast clouds of dust and gas into their tails. Nineteenth-century skies were graced with more sungrazers than have been seen so far in the 20th century, a fact that has led astronomers to conclude that families of comets travel in loose bunches. Strengthening this scenario is the fact that groups of sungrazers seem to have virtually the same orbit. The Great Comet of 1882 and Ikeya-Seki in 1965 followed paths almost identical to that of the comet of 1106.

The Great Comet of 1843 — a sungrazer — is believed to have had the longest tail ever recorded. One observer reported that it stretched over 90 degrees of the sky and was equal to or greater than the brightness of the Milky Way. It spanned 200 million miles — more than twice the distance from Earth to the sun. But even more incredible was the fact that the comet survived a sweep that carried it a mere 80,000 miles above the sun's surface — one-third the distance from Earth to the moon. If the sun had not been at minimum activity, the comet might have been snuffed out by a solar prominence.

Visually, comets have an almost starlike core surrounded by a misty haze called the coma. The diaphanous tail, which sweeps back from the coma, is usually noticeably fainter than the coma and shows little structure. Occasionally, delicate lengthwise striations make the tail look like strands of hair or give it a subtle feathery appearance. The word comet is derived from the ancient Greek word *kometes*, which means "wearing long hair."

Either with the unaided eye or by telescope, the comet will appear yellowish white. Colour photographs record not only a yellow dust tail but also a bluish tail with a wispy veil-like structure called the gas tail, which is much more difficult for the eye to discern. As is the case with nebulas, the hues of comets are almost completely invisible to the eye. To be spectacular, a comet must be bright *and* have a large dusty tail. Fainter comets are often tail-less, appearing simply as a concentrated cloud of nebulous material that gradually fades off toward the edges. No matter how bright the nucleus and coma, a comet is a relatively small object in the sky and difficult to distinguish from a star with the unaided eye.

Comets vary greatly in the shape and length of the dust tail and in the brightness of the coma. Some have bright comas and faint stubby tails, some have fanlike tails, and others eject long pencil-thin tails. Frequently, the tail is curved, due to the comet's motion in its orbit. The gas tail, however, is always straight, since it is directly influenced by the solar wind, electrically charged particles from the sun.

In my experience, comets below eighth magnitude are unimpressive; the excitement is in finding them. When located, they look like faint globular star clusters. Of course, the challenge in finding a comet is the same as that in finding a galaxy or a star cluster of similar magnitude. And comets do change position fairly rapidly from night to night.

Sometimes, this motion can be detected. In May 1983, Comet IRAS-Iraki-Alcock came within 16 times the moon's distance from Earth, closer than any comet since Lexell in 1770. At the time of its closest approach, the comet was a third-magnitude blob about the size of the full moon. I set up my telescope, trained it on the comet and started to watch. Within seconds, I noticed that it was moving. It was actually possible to see the comet plying its way through the inner solar system.

Some comets have violent surges in brightness, perhaps as their weakened structure breaks apart and volatile ices are exposed to solar radiation. In 1973, a faint comet named Tuttle-Giacobini-Kresak increased in brightness 10,000 times over a period of just a few nights, then subsided back to its original level.

In general, when a comet rises above fifth magnitude, usually when it is within Mars's orbit, every effort should be made to observe it each clear night (or early morning, since half the comets appear in the morning sky). Dark skies substantially enhance the detail visible in a comet's delicate tail structure – especially its overall length. Low-power binoculars give the best views of brighter comets, since the tails usually extend several degrees. A comet of second magnitude or brighter with a good tail is something seen once or twice in a lifetime.

While bright comets are always reported in the press, the stories seldom describe where to look for them. Many astronomy clubs provide a notification service for moderately bright comets, and they sometimes have telephone alerts for unusual, rare celestial phenomena. The most recent bright naked-eye comet visible from mid-northern latitudes was Comet West in 1976, but hardly anyone, apart from astronomy enthusiasts, saw it. There was abysmally poor newspaper publicity about West because editors had been burned by the overblown predictions for the lacklustre Comet Kohoutek two years earlier. Comet brightness is the one factor that still defies precise predictions because of the uncertainties in the composition and melting rate of the comet's nucleus.

Telescopically, a comet nucleus is generally seen as a starlike point, but it is sometimes so shrouded by the coma that it appears simply as a brightened concentration in a haze – like a distant streetlight in mist. Frequently, no structure is detected, especially in fainter comets. The brightest comets show the most activity, and the structural details around their comas are often seen better visually than in photographs. Just as interesting as the specific details of the comet are the changes in size, shape and brightness from night to night as it slowly shifts its position against the starry background.

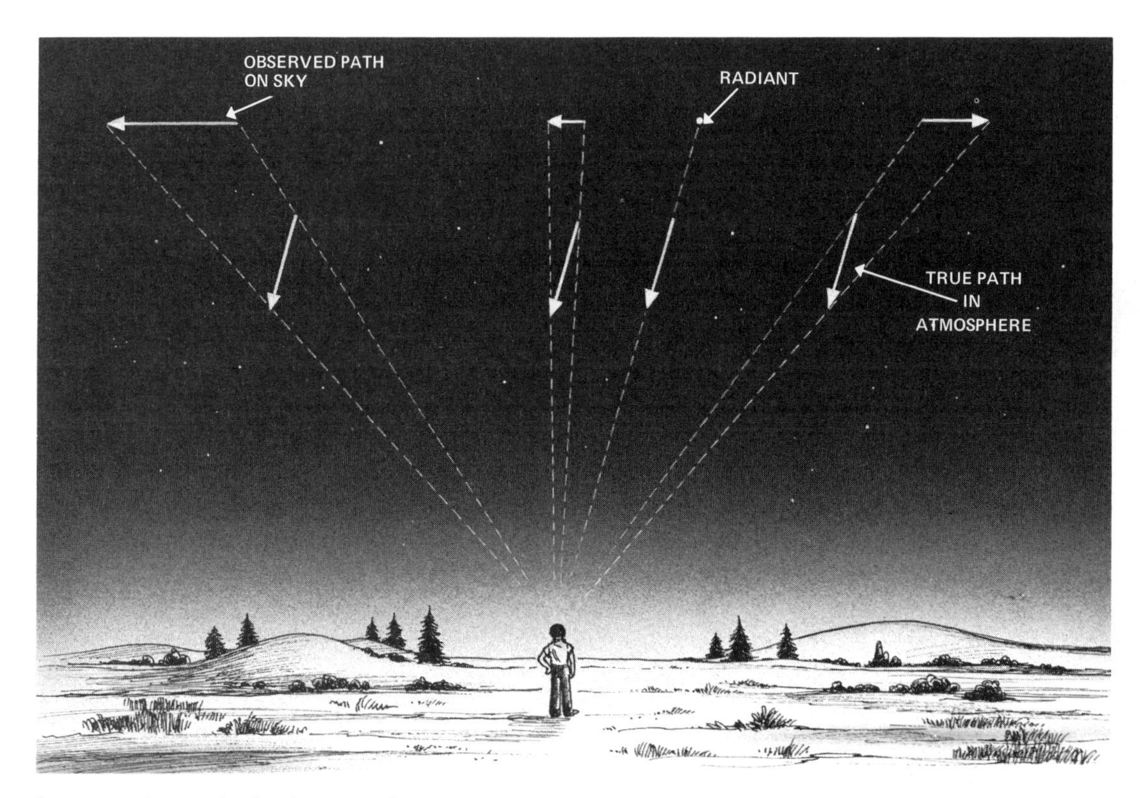

During a meteor shower, the meteors appear to emerge from a point called the radiant.

METEORS

It happens to all of us at one time or another: one quick glance up at the star-filled sky — perhaps for just a second — and suddenly, the placid view is sliced by the brilliance of a falling star. The expression "falling star" is just a description of what appears to be occurring; the object that flashes across the sky and quickly disappears is properly called a meteor.

Meteors have nothing to do with stars. They are tiny bits of space debris so small that thousands would easily fit in the palm of your hand. Yet each one of them causes that familiar brief but brilliant flash in the night sky, a dazzling flameout as it ends its existence in a 40,000-mile-per-hour plunge into the upper layers of the Earth's atmosphere. At such velocities, friction with air particles vaporizes an average-sized meteor in less than a second. It is the vaporization process that produces the sudden flash of light.

Each day, Earth collects about 400 tons of meteoric debris, most of it microscopic dust so small that it does not produce visible meteors but merely collides with the Earth's atmosphere and floats to the ground months or years later. A minority of the impacting pieces are big enough to flash as visible meteors. On rare occasions, a chunk large enough to survive the fiery plunge hits the ground as a meteorite.

Meteor terminology is cumbersome and confusing, but some of the words need definition. A *meteor* is the bright streak of light seen at night, as a small bit of space debris burns up in the Earth's atmosphere. A *meteoroid* is a small chunk of matter in space that could become a meteor. A *meteorite* is a piece that survives its descent through the Earth's atmosphere and reaches the surface. The related term *asteroid* refers to small subplanetary bodies that orbit the sun, generally between the orbits of Mars and Jupiter. Large meteorites are believed to be de-

A Perseid meteor slashes the sky during a 10-minute time-exposure photograph.

bris from collisions between asteroids.

Regardless of the confusing terms, the objects seen in the night sky are meteors. On an average night, three or four moderately bright meteors per hour can be seen, with the rate rising to seven or eight per hour by dawn. However, as Earth navigates its annual orbit around the sun, it encounters swarms of meteoric material at predictable intervals. Located at specific points in the Earth's orbit, like mileage posts on a racetrack, the meteor swarms are rapidly encountered and soon left behind. The result is a meteor shower lasting a few nights at most. The best meteor showers can produce one meteor per minute on average, although most yield fewer than that.

Whether a meteor shower is occurring or not, the peak period for meteors is from about 1 a.m. to dawn. The reason for this is that after midnight, the nighttime side of Earth faces in the direction in which it is moving in its orbit around the sun. The after-midnight, or "forward," side of Earth sweeps up more meteors than the before-midnight, or "trailing," side. As an analogy, when I am out walking during a heavy snowfall, the front of my coat becomes plastered with snow, while my back is only slightly peppered with flakes. This, of course, is due to forward motion — I walked into the flakes. Earth does the same as it pursues its orbit at a constant 67,000 miles per hour.

Experienced meteor watchers use lawn chairs that adjust to a nearly horizontal position so that as much of the sky as possible is comfortably in view. Select the darkest available site, and face in the direction of the meteor radiant (see table). Remember to have a good supply of blankets and insect repellent. Binoculars or telescopes are useless for observing meteors because their fields of view are far smaller than that of the human eye. Meteors can dart unpredictably from practically anywhere. Since meteor observing requires no telescopic equipment, it is a unique opportunity to introduce astronomy to others and to become reacquainted with the stars and constellations.

Serious meteor watchers record each meteor's track on a photocopy of an all-sky chart, such as those in this book. This requires a marking pen and a heavily filtered flashlight so that night vision will not be spoiled. When a meteor is spotted, its starting and ending points among the stars are carefully noted, then the path is plotted on the chart, with an arrow indicating the direction of flight. Tracing the lines backward at the end of the night reveals the radiant point of the meteor shower. A few meteors will not radiate from this area. These are the random non-shower meteors seen every night.

Meteors from the recognized showers are known to be debris from comets. The comets relinquish the meteoric material during close encounters with the sun. Their icy bodies partly vaporize, releasing embedded dust and denser bits of ice, which spread along the comet's orbit like a trail of sand from a punctured sandbag. When Earth crosses this trail of debris, a meteor shower occurs.

Even during the most abundant of the regular meteor showers, when several dozen meteors per hour may be observed under favourable circumstances, the average distance between individual meteors is more than 100 miles, so the "swarm" that produces the shower is, in fact, almost empty space. No meteor belonging to any meteor shower has ever been known to reach the Earth's surface. Even those as bright as Venus are quickly consumed, but they often leave a glowing trail that may last for several seconds.

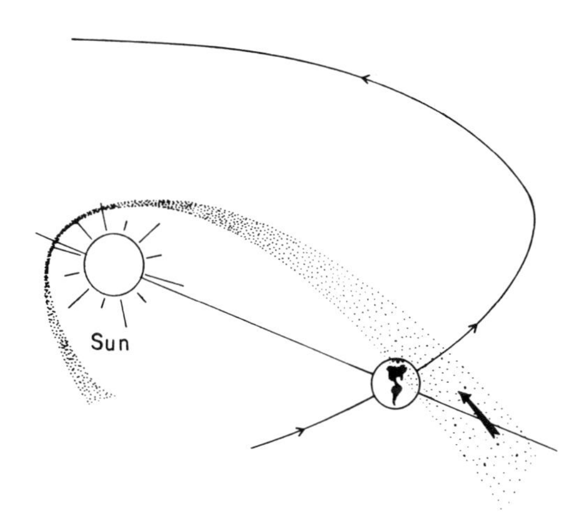

Meteor showers occur the same nights each year, as Earth passes through cometary debris.

AURORAS

Diaphanous curtains of green, white and red dance in the northern sky, billowing and swirling as if propelled by some distant cosmic wind. Most of us have seen this nocturnal spectacle at one time or another — the aurora borealis, or northern lights. Some nights, they look more like pulsing clouds or filmy arcs of light on the northern horizon. Or, more rarely, the sky is alive with roiling luminescence whirling overhead.

MAJOR ANNUAL METEOR SHOWERS

NAME OF SHOWER	RADIANT	DATE OF MAXIMUM	HOURLY RATE AT MAXIMUM*
Quadrantids	NE (Draco)	Jan. 4	25–100
Lyrids	NE (Lyra)	Apr. 21	5–15
Eta Aquarids	E (Aquarius)	May 4	10–20
S. Delta Aquarids	SE (Aquarius)	Jul. 27-29	10–35
Perseids	NE (Perseus)	Aug. 12	30–70
Orionids	E (Orion)	Oct. 20	10–30
S. Taurids	E (Taurus)	Nov. 3-7	5–15
Leonids	E (Leo)	Nov. 16	10–20
Geminids	E (Gemini)	Dec. 13	20–80
Ursids	N (Ursa Minor)	Dec. 22	5–15

*The range of values reflects the variations in the strength of the meteor displays from year to year. These figures do not include the half-dozen or so sporadic meteors seen each hour.

Canada is the best place in the world for aurora watching. From southern British Columbia and southern Ontario, more than a dozen auroral displays are seen annually on the average. Elsewhere in the country, the odds are even more favourable, reaching a maximum of one every third or fourth night from the Yukon, the Northwest Territories, northern Quebec and Labrador. But auroral visibility is not confined to northerly latitudes. Several times a year, auroras are seen from the northern half of the United States. And about once every two or three years, auroral lights dance in Florida's skies.

The most persistent (and entirely wrong) explanation for the phenomenon is sunlight reflecting off polar ice. As long ago as the early 1900s, Norwegian astrophysicist Carl Stormer correctly explained that the aurora originates with energetic particles blasted in the Earth's direction by eruptions on the sun. Travelling at speeds of millions of miles per hour, these particles reach Earth in a day or two, but instead of plunging directly into the atmosphere, they are deflected by the planet's magnetic field. The particles then funnel along the magnetic lines of force, eventually entering the atmosphere over the magnetic poles, where interaction with upper-atmosphere molecules produces the aurora borealis and aurora australis — the northern and southern lights.

Trillions of watts of power are pumped into the upper atmosphere during an auroral display, creating not only the visible light but also heat, ionized particles and even some intense (though far from dangerous) radiation. The aurora must affect the Earth's weather machine, but exactly how or in what manner is unknown.

Records of aurora borealis observations can be traced as far back as the sixth century B.C. Chinese scholars described the lights as a fire dragon among the stars. The most abundant auroral displays recorded in China occurred in the 11th and 12th centuries A.D., coinciding with recently discovered high levels of solar activity. Greek accounts by Anaximenes and others leave no doubt that the phenomenon has always been a source of wonder.

The great light show is generated when the high-energy electrons and protons emitted by solar flares are directed by the Earth's magnetic field lines into collision courses with the upper atmosphere. Atmospheric atoms and molecules hit by the inflowing particles become ionized by the impacts (electrons are knocked off), thereby absorbing some of the energy of the impinging particles. The ionized atoms and molecules soon return to a stable state by releasing the absorbed energy, often in the form of visible light. Auroral displays are a combination of these interactions at altitudes ranging from 60 to 600 miles, although the most sensational auroras occur about 70 miles above Earth, where the atmosphere is at the right density to produce the maximum effect.

Auroral colours are caused by the different types of atoms stimulated in this way. Oxygen emits a greenish white light or a red hue at extremely low atmospheric pressures; ionized molecular nitrogen produces a bluish tinge.

Because of the direct relationship between auroras and solar activity, periods of high solar turbulence near sunspot maximum always produce increased auroral activity. At sunspot maximum, auroras are more likely to be seen farther south because the magnetosphere — the cometlike bubble of magnetically trapped particles around Earth — is stretched, and the auroral zone enlarges, moving toward the equator enough to allow millions more people to see the dancing lights.

A typical auroral display begins with a white or pale greenish glow toward the northern horizon. Then a few spikes or arcs slowly crawl up the sky, brightening as they rise. As the aurora increases in intensity, vertical bands begin to pulsate and waver, developing into waves and folds that form fragile celestial draperies. In the best displays, pulsating curtains in hues of red, green and white can last from a few minutes to hours.

In an intense display, the undulating curtains are overpowered by swirling, billowing clouds that lurch over the sky in a matter of seconds. The very best auroras develop into brilliant streamers radiating from overhead, and the sky fills with light and colour, obliterating almost all the stars in lashing, pulsing fury. Called coronal auroras, these are one of nature's great spectacles. During a coronal aurora, the rays are actually parallel, but they appear to converge at the magnetic zenith, just as railroad tracks seem to converge in the distance. The magnetic zenith for most of the populated parts of Canada and most of the United States is slightly south of the overhead point. To be fully appreciated, auroras must be observed from a dark location. Their delicate structure and hues are wiped out by streetlights, backyard illumination or even the glow from urban lighting in general.

Reports that people are able to hear auroras are so persistent that some scientists are becoming convinced the phenomenon is real. There are hundreds of accounts of crackling, hissing sounds associated with auroras, but only some people hear them. In a group, usually one person notes the sound. The origin of such sounds is still debated, but it illustrates the fact that cosmic mysteries are as near as the top of the Earth's atmosphere.

A vigorous multihued auroral display can be ranked second only to a total solar eclipse among nature's sky spectaculars. Occasionally, brilliant red curtains, **right**, *overpower the more common green celestial draperies. This display, one of the best of the 20th century, was seen over much of North America the nights of March 12 and 13, 1989. Sometimes, green, yellow and pink mingle,* **left**, *but usually, white and green dominate,* **above**. *Auroras do not occur more frequently at any particular time of year or hour of night. However, the displays are much more common in the months around sunspot maximum.*

11

Photographing the Night Sky

Astronomy offers one of those pleasures which follows the law of increasing, rather than diminishing, returns. The more you develop it, the more you enjoy it.

— Viscount Grey

All of the photographs in this book, except for a few spacecraft images of the planets, were taken by amateur astronomers. Some of the pictures were simple to obtain; others required specialized equipment, hours of preparation and plenty of skill and experience on the part of the photographer.

Surprisingly good pictures of the natural beauty of the nighttime sky are possible with a standard 35mm camera. This chapter is a step-by-step guide to the easiest types of astrophotography.

First, the camera. Since the vast majority of successful astrophotographs are taken with 35mm single-lens-reflex (SLR) cameras, you really should have one even for dabbling in celestial photography. The camera must have a feature that allows the shutter to remain open for at least 12 seconds — preferably indefinitely. Some inexpensive cameras, and a few very expensive ones, do not have this option. Look for a setting marked "B" (bulb), or check the owner's manual for time-exposure instructions.

Until recently, all 35mm SLR cameras had time-exposure (bulb) settings, but some of the newer models with built-in computers, which do everything except compose the shot, cannot be told that you don't want them to do anything. A plain old manual 35mm camera like my Olympus OM-1 is best. There are no batteries that will drain during a time exposure and no computers to terminate my purposely overexposed shot prematurely.

Next, the film. Use colour. Black and white films have no superiority except in advanced applications well beyond what beginners should try. There is more colour in the sky than the eye can see, and modern colour films pick it up easily. This is one area where recent advances work completely to the benefit of amateur astronomers. New films like Fujichrome 1600D (for slides) and Konica SR-V 3200 and Kodak Ektar 1000 (for prints) are vastly improved for night photography over anything available previously. These are my favourite films, but any film rated at ISO 400 or higher will work well.

With slide film, your "negative" is your slide. What you see projected on your home screen is what you shot. If you goofed, it is apparent. Colour print films are different in that the rectangular picture you get back from the photofinisher is a printed version of the negative.

Photofinishers (whether machines or humans) are used to printing pictures of children, weddings and vacation scenery. Astronomical photographs often throw them. Sky backgrounds can appear anywhere from black to green to pink. Seldom do you get the best print that can be made from your negative. What to do? Tell your photofinisher that there are "night-sky pictures with stars" on the film. This alerts the lab to something unusual and reduces the chances of wretched prints.

If you think you have captured something good, get a custom reprint. Although these prints cost more, you can specify exactly what you want, and they are obliged to do it again and again, if necessary, to get it right. Always take a sample of good astrophotography to give

them an idea of what you are looking for.

Most 35mm cameras have interchangeable lenses. This allows both wide-angle and telephoto possibilities. Zoom lenses have become popular in recent years. If zooms are all you have, use them at, or near, minimum focal length (e.g., 28mm to 40mm for a 28mm-70mm zoom). However, fixed-focal-length lenses are preferred for astronomical photography because they have "faster" focal ratios and usually produce sharper images. Now, on with how to take a celestial photograph.

For the easiest type of sky photography, you need a 35mm camera with a 55mm (or shorter) focal-length lens, tripod and cable shutter release. The cable release (less than $10) is necessary to lock the shutter open for time exposures. Mount your 35mm camera, loaded with ISO 400 or faster film, on the tripod, turn off anything automatic that can be turned off (light meter, flash, et cetera), and set the lens at f/2.8, f/3.5 or f/4. Focus at infinity. Set the shutter speed at B, and aim the camera at a familiar stellar configuration, say, Orion or the Big Dipper. Place the lens cap over the lens. Press the cable release, and lock the shutter open with the lock on the cable.

Now you are ready for the picture. Remove the lens cover for 12 to 15 seconds. Replace the cover, and release the cable lock. Using this technique with the standard 50mm or 55mm lens, you will get realistic pictures of the night sky. Double the exposure with lenses in the 28mm range. Choose a moonless night away from the city's glow for best results. Stars

Star trails, **previous page**, *are easy to capture and show real star colours. Kodachrome 200 film was used for the 90-minute exposure with a 24mm lens at f/3.5. A much shorter 15-second exposure on Konica 3200 film,* **right**, *recorded stars down to eighth magnitude in the Big Dipper region. A 50mm f/2.8 lens was used. Photographs like those on pages 19 and 153 can be obtained with a piggyback setup,* **left**. *For photography through the telescope, use a T-ring, 1¼-inch adapter or 2-inch adapter, shown left to right,* **above**. *The telescope then acts as a telephoto lens.*

well below the naked-eye limit will be recorded on ISO 1000 or faster film. Exposures longer than the maximum suggested will produce trails instead of dots for stars.

The Earth's rotation not only makes the sun appear to move across the sky from east to west but also imparts a motion on the stars. Capturing this motion on film produces impressive star trails, which reflect the rotation of Earth precisely. Stars above the North Pole appear to move hardly at all, just as a point near the pole on the Earth's surface barely moves compared with a point on the equator.

Set up the camera as described above, but leave the shutter open for at least 15 minutes to record the motion of the Earth's rotation. Photographs taken with the camera pointing toward the southern sky or overhead will show long east-to-west trails, whereas photographs taken pointing toward the north pole of the sky (an imaginary point very close to the north star, Polaris) produce curved star trails with the geometrical centre of each curving arc at the north celestial pole. A photograph of an hour or more will show beautiful swirling trails, as the stars appear to pivot around Polaris.

The star colours recorded by colour film are true colours. Each star is a slightly different temperature. Hot stars are blue, cooler stars white, and even cooler stars are shades of yellow and orange.

Occasionally, an aircraft or satellite will sweep across the camera's field of view during star-trail photography. In both cases, the track on the photograph

Portrait of the North America Nebula, **right**, *taken from the author's driveway, was a four-minute exposure with a 90mm lens at f/4 using Konica 3200 film. The camera rode piggyback on an equatorial-mounted telescope. Moon photograph,* **above**, *was taken through a beginners' 60mm refractor on Kodak Technical Pan film. Similar lunar shots are possible with ordinary snapshot colour film.*

will be straight, but the aircraft's flashing lights may produce an interesting beads-on-a-string effect. Another unpredictable intruder, and by far the most prized unintentional catch, is a meteor slashing across the star trails. Only bright meteors will show up; the faint ones are too feeble to register on film during their brief flights.

Try meteor hunting with a camera by setting up for star-trail photographs on the peak nights of the best showers, such as the Perseids or Geminids. Aim the camera in the general direction of the radiant (see Chapter 10), but not right at it. It is largely a matter of luck, but some amateur astronomers have captured some remarkable sights using this method.

A telephoto lens of 100mm focal length or longer expands the repertoire of available celestial objects to include the moon. Some of the most beautiful pictures of our satellite are taken just before the sky is completely dark. The moon forms a backdrop for trees or houses or other foreground objects. (Lenses of less than 100mm focal length do not show the moon large enough to distinguish any of the darker surface areas that characterize its features.) Telephoto shots of the moon are best regarded as trial-and-error efforts, although modern light-metering systems can be fairly accurate. Exposures are generally less than one second, sometimes as little as 1/250 of a second, but the important point is to take a range of exposures and to record the pertinent data so that successes can be repeated. A camera tripod is usually necessary.

A rewarding and relatively uncomplicated extension of naked-camera photography is piggyback-guided photography. Simply mount an SLR camera on an equatorially mounted telescope. Most major telescope manufacturers have adapters for this purpose. The telescope now becomes a guider for the camera. Watch a selected guide star through the main telescope, and keep it centred during the exposure using the telescope's slow-motion controls. (This type of photography is almost impossible without slow-motion controls in both axes.) Star trailing is eliminated because the telescope keeps the star field in a fixed position for the camera, thus allowing longer exposure times and the recording of much fainter stars. During the time exposure, the stars will remain as points. Use wide-angle or normal-focal-length lenses, at least initially, since accurate tracking of the stars is not as critical as it is with telephoto lenses.

A complete guide to photography using a telescope would require far more detail than I can go into here. Furthermore, like telescope making, astrophotography *through* a telescope is practised seriously by only a small minority of backyard astronomers. There are several reasons for this. It demands a great deal of patience, plenty of trial and error and, for most types of telescopic astrophotography, very dark skies. In some cases, it also requires elaborate equipment. However, it *is* possible for a beginner to take some fine photographs using a camera in conjunction with a telescope.

A relatively inexpensive (about $30) camera/telescope adapter is a necessity. It replaces the lens on an SLR camera with a tubular holder the same size as a telescope eyepiece. The camera then replaces the eyepiece and uses the telescope as a telephoto lens. This method is almost foolproof for lunar photography and also works well on terrestrial objects. I have photographed birds hundreds of feet away as if they were within reach. To photograph the sun, the same filters are needed as for visual observation.

Photographing the planets is *much* more difficult. For the most part, the results are disappointing, because even the best planetary photographs never show as much detail as can be seen visually through the eyepiece of the same telescope. Planetary photography also requires additional adapters to place an eyepiece ahead of the camera in order to boost the image to a suitable size. Planet photographs, or extreme close-ups of lunar features, need exposures of up to 15 seconds.

The final challenge for the astrophotographer is the unlimited number of galaxies, nebulas and star clusters beyond the solar system. To attempt to shoot them properly, an accessory called an off-axis guider is needed, because recording faint objects requires time exposures of at least several minutes. The telescope has to remain precisely aimed during the exposure. This involves motor-driven slow-motion controls on the telescope and a push-button control paddle to keep a guide star centred while shooting the main object. To do it right requires several hundred dollars' worth of accessories and more instructions than can be provided here.

Because the field of astrophotography has been almost completely revolutionized in the last decade by high-speed films, hypersensitizing and dry-ice cold-camera photography, to name just a few techniques, up-to-date references are elusive. In *The Backyard Astronomer's Guide*, the companion volume to *NightWatch*, three chapters focus on astrophotography, offering a fairly comprehensive reference to this subject. For more information, check the astrophotography articles in *Sky & Telescope* and *Astronomy* magazines over the past few years.

Many of the photographs in this book were taken by amateur astronomers who have become so proficient that their work rivals photographs taken with the largest professional telescopes, an amazing testimony to their dedication and ingenuity.

12

Resources

What is it all but a trouble of ants in the gleam of a million million suns?

— Alfred, Lord Tennyson

All leisure interests require a certain minimum investment in equipment and supplies. Apart from a telescope, you will need good binoculars, a few general astronomy books, one or two practical guidebooks, an annual data-reference book and a subscription to at least one astronomy magazine.

MAGAZINES

There are two major monthly magazines designed for, and in many instances written by, amateur astronomers. Thousands of astronomy enthusiasts subscribe to both, and reading at least one of the two regularly will keep the enthusiast up to date on celestial events and discoveries. In addition, the two magazines are the only place to find out who sells what astronomy equipment. Both are filled with informative ads.

Sky & Telescope, Box 9111, Belmont, MA 02178. This is the oldest, most thorough and most distinguished astronomy magazine. All serious amateur astronomers, many professionals and thousands of libraries subscribe to it. Copies are found on larger newsstands, and it is often sold by planetarium shops and telescope dealers. The publication attempts to cover the entire field from recent discoveries in astrophysics to reports on amateur-astronomy conventions. Current sky events are well covered, with excellent diagrams and charts.

Astronomy, Box 1612, Waukesha, WI 53187. Established in 1973, this publication is the world's largest-circulation English-language astronomy magazine. Most well-stocked newsstands carry it. Aimed at the beginning amateur astronomer, it is definitely easier reading than *Sky & Telescope*. Each issue contains one or two lavishly illustrated features along with articles on the use of astronomical equipment, techniques of observation and current celestial events.

Mercury, the bimonthly magazine of the Astronomical Society of the Pacific (ASP), 390 Ashton Avenue, San Francisco, CA 94112, is available only through membership in the society. In addition to *Mercury*, ASP members receive the *Abrams Planetarium Sky Calendar*, described below, for observers throughout North America.

The Planetary Report, a bimonthly magazine devoted entirely to planetary exploration and the search for extraterrestrial life, is available through membership in The Planetary Society, 65 N. Catalina Avenue, Pasadena, CA 91106.

Other magazines of special interest to amateur astronomers include *The Observer's Guide*, published bimonthly by Astro Cards, Box 35, Natrona Heights, PA 15065, and the *Griffith Observer*, a small monthly published by the Griffith Observatory, 2800 E. Observatory Road, Los Angeles, CA 90027.

The monthly *Abrams Planetarium Sky Calendar* is a concise digest of what is visible with unaided eyes and binoculars. It is not a magazine, just two sides of a standard-sized page per month, but a surprising amount of information is clearly presented in neat diagrams that are useful to beginners and veteran astronomers alike. A year's subscription is just a few dollars. For subscription information, write: Sky Calendar, Abrams Planetarium, Michigan State University, East Lansing, MI 48824.

GENERAL ASTRONOMY BOOKS

In recent years, the best overviews of current astronomical knowledge have been provided by college textbooks such as William Hartmann's *Astronomy: The Cosmic Journey* (Wadsworth Publishing; Belmont, California) and *Astronomy: From the Earth to the Universe* by Jay M. Pasachoff (Saunders; Philadelphia). Both contain hardly any mathematics and are well written, copiously illustrated and updated and revised every few years. The best encyclopaedic works that I have seen are *The Cambridge Atlas of Astronomy* (Cambridge University Press; New York; revised edition 1988) and *The International Encyclopedia of Astronomy* (Crown; New York; 1987).

Less technical than any of the above is *The Universe and Beyond* by Terence Dickinson (Camden House; Camden East, Ontario; revised edition 1992). It is packed with full-colour photographs, illustrations and diagrams, more than half of them never published in book form before. Emphasis is on the full spectrum of astronomical enquiry from solar system exploration to cosmology to extraterrestrial life. It also contains complementary tables to those in *NightWatch* and has an extensive guide to further reading.

BACKYARD ASTRONOMY GUIDES

For many years, I have wondered when the definitive up-to-date guide for amateur astronomers would appear — one that would cover in detail the choices in equipment and techniques for practising backyard astronomy. In 1986, I asked Alan Dyer, now associate editor of *Astronomy* magazine, if he would be interested in collaborating on such a book. Five years later, *The Backyard Astronomer's Guide* was published by Camden House. It has received the same warm reception as *NightWatch* and is already in its second printing. If you have found *NightWatch* a suitable guide to amateur astronomy, I invite you to take the next step to *The Backyard Astronomer's Guide*.

A classic reference that I highly recommend is *Burnham's Celestial Handbook* by Robert Burnham (Dover; New York; 1978), a colossal 2,100-page three-volume set that took literally decades to prepare. In softcover, the three books are about $20 each. Every significant multiple star, variable star, star cluster, galaxy and nebula visible in amateur astronomers' telescopes is described or tabulated in detail.

A practical guide for backyard astronomers is the *Mag 6 Star Atlas* by Terence Dickinson, Victor Costanzo and Glenn Chaple (Edmund Scientific; Barrington, New Jersey; 1982). Charts and tables of interesting objects face each other for quick reference. Many pages of useful information on telescopes and observing round out this softcover guidebook. Price: about $20.

Another favourite sixth-magnitude atlas is *Bright Star Atlas* by Wil Tirion, published by Willmann-Bell Inc., Box 35025, Richmond, VA 23235 (write for excellent free catalogue). All stars to magnitude 6.5 are shown along with thousands of deep-sky objects and reference tables. A bargain at $10. A colour version ($20) using charts in a slightly different format is *The Cambridge Star Atlas 2000* by Wil Tirion (Cambridge; New York).

Sky Atlas 2000 by Wil Tirion (Sky Publishing; Cambridge, Massachusetts; 1982), a well-designed, beautifully detailed eighth-magnitude-limit star atlas, is available in either full-colour ($40) or black-and-white ($20) versions. This atlas has quickly become a standard reference for serious backyard astronomers. Tables of information on the thousands of objects plotted in the atlas are available as supplementary books from Sky Publishing.

The ultimate star atlas for amateur astronomers is *Uranometria 2000*, which shows stars to ninth magnitude — 330,000 of them on 473 charts bound in two rugged 9-by-12-inch volumes (about $100 a set). Virtually every deep-sky object visible in any amateur telescope is plotted and named. I cannot imagine this atlas being surpassed until well into the 21st century. It is not for beginners, but all experienced stargazers will want to own this definitive work. Published by Willmann-Bell Inc., Box 35025, Richmond, VA 23235.

All About Telescopes by Sam Brown (Edmund Scientific; Barrington, New Jersey; 1967). This 192-page book, with more than 1,000 illustrations, was the outstanding telescope-making reference for many years. Although badly in need of revision, it is still a classic. The best book for the average first-time telescope maker is Richard Berry's *Build Your Own Telescope* (Scribner's; New York; 1985). Berry is an expert designer of simple, rugged backyard telescopes.

A good starting point for beginning astrophotographers is *Astrophotography Basics*, a brief but otherwise excellent booklet on the subject, available free from Kodak, Customer Service Dept., Rochester, NY 14650, or Kodak Canada Inc., Publications Dept., 3500 Eglinton Avenue West, Toronto, ON M6M 1V3. Another useful reference is *Astrophotography for the Amateur* by Michael Covington (Cambridge University Press; New York; 1985).

The Astronomical Companion by Guy Ottewell (published in 1979 by the Department of Physics, Furman University, Greenville, SC 29613) successfully melds many concepts that are left dangling in introductory astronomy books. Ottewell has crafted a series of brilliant sketches that puts celestial phenomena and the large-scale structure of the universe into perspective. It makes an ideal follow-up to *NightWatch*. I highly recommend it. Price: $12, softcover.

One of the finest books ever written about the romance of amateur astronomy is *Starlight Nights* (Sky Publishing; Cambridge, Massachusetts; 1980), the autobiography of the late Leslie Peltier, probably the foremost nonprofessional astronomer of the 20th century. This sensitive and insightful narrative is one of the most memorable books on the subject.

COMPUTER SOFTWARE

The evolution of astronomy programs for personal computers has been swift and dramatic in the past few years. For IBMs and compatibles, I can suggest *Dance of the Planets* by ARC Software, Box 1955, Loveland, CO 80539, a powerful program that is especially strong in simulating views of solar system objects from user-selected positions in space. More suitable for beginners because of its general content is *EZCosmos*, Future Trends Software, Suite 102, 1601 Osprey Drive, DeSoto, TX 75115. Another general program is *The Sky for Windows*, Software Bisque, 912 12th Street, Golden, CO 80401. All three are reasonably user-friendly.

But my favourite astronomy software is, as of this writing (early 1992), available only for Macintosh and Amiga computers. Called *Voyager*, it is awesome in its capabilities as a home planetarium. *Voyager* is available from Carina Software, 830 Williams Street, San Leandro, CA 94577.

TELESCOPE EQUIPMENT & ACCESSORIES

The following list includes many of the leading manufacturers of astronomy equipment and their products. Most of these companies will supply a product catalogue upon request. This is *not* a listing of *dealers*. Many dealers sell all brands. For the dealer nearest you, check the Yellow Pages under Telescopes or contact the manufacturer for a dealer list.

Astro-Physics
11250 Forest Hills Road
Rockford, IL 61111

Apochromatic refractor telescope specialist; heavy-duty equatorial mounts, accessories.

Edward R. Byers Company
29001 W. Hwy. 58
Barstow, CA 92311

Precision equatorial mountings.

Celestron International
2835 Columbia Street
Torrance, CA 90503

Schmidt-Cassegrain telescopes, refractors, Newtonian reflectors, binoculars, many accessories.

Coulter Optical Company
Box K
Idyllwild, CA 92349

Dobsonian-type Newtonian reflectors.

DayStar Filter Corporation
Box 1290
Pomona, CA 91769

Precision solar filters, nebula filters.

Edmund Scientific Company
101 E. Gloucester Pike
Barrington, NJ 08007

This large general-science supply house stocks popular, inexpensive telescopes and accessories.

Jim's Mobile Inc.
1960 County Road 23
Evergreen, CO 80439

Motorized focus controls, motors for telescope mounts, large-aperture Newtonians.

Lumicon
2111 Research Drive, #5
Livermore, CA 94550

Specialists in telescope accessories, particularly photographic gear.

Meade Instruments Corporation
1675 Toronto Way
Costa Mesa, CA 92626

Schmidt-Cassegrain telescopes, refractors and Newtonian reflectors; accessories for all types of telescopes.

Kenneth F. Novak & Company
Box 69
Ladysmith, WI 54848
Accessories and parts for Newtonian reflectors.

Orion Telescope Center
2450 17th Avenue
Santa Cruz, CA 95061
Telescope and observing accessories.

Parks Optical
270 Easy Street
Simi Valley, CA 93065
Newtonian reflector parts, eyepieces, accessories, complete telescopes.

Questar Corporation
Route 202, Box 59
New Hope, PA 18938
Premium Maksutov-Cassegrain specialists.

Sky Designs
4100 Felps #C
Colleyville, TX 76034
Large-aperture Dobsonian reflectors.

Spectra Astro Systems
6631 Wilbur Avenue, Suite 30
Reseda, CA 91335
Astrophotographic accessories.

Tele Vue Optics, Inc.
20 Dexter Plaza
Pearl River, NY 10965
Refractors and eyepieces; wide range of high-performance oculars.

Thousand Oaks Optical
Box 248098
Farmington, MI 48332
Solar filter specialist.

Roger W. Tuthill, Inc.
Box 1086
Mountainside, NJ 07092
Mylar solar filters and other accessories.

University Optics, Inc.
Box 1205
Ann Arbor, MI 48106
Extensive line of eyepieces and accessories.

SLIDES, POSTERS & CHARTS

Both *Astronomy* and *Sky & Telescope* offer selections of posters and slides through their own sales departments. Most major observatories do not sell slides and posters directly to the public. However, many of their best photographs are available in slide and poster form from Hansen Planetarium Publications, 1098 S. 200 West, Salt Lake City, UT 84101. Astronomical slides and posters are also available from Edmund Scientific Company, 101 E. Gloucester Pike, Barrington, NJ 08007; Astronomical Society of the Pacific (see *Mercury*, page 154); The Planetary Society, 65 N. Catalina Avenue, Pasadena, CA 91106; and Everything in the Universe, 5248 Lawton Avenue, Oakland, CA 94618.

ANNUAL PUBLICATIONS

Observer's Handbook, published annually by the Royal Astronomical Society of Canada, 136 Dupont Street, Toronto, ON M5R 1V2, is a mandatory reference for any backyard astronomer. Dozens of tables for celestial phenomena range from times of moonrise and moonset to the positions of Jupiter's moons and the location of meteorite scars in North America. Thoroughness and accuracy have made the *Observer's Handbook* the most widely used annual reference for amateur astronomers.

Astronomical Calendar by Guy Ottewell is published by the Department of Physics, Furman University, Greenville, SC 29613. Write to the author for a descriptive brochure and for the current price of this excellent annual guidebook. Ottewell's unique illustrations and emphasis on diagrams, rather than tables, produce little direct overlap with the *Observer's Handbook*.

AMATEUR ASTRONOMY CLUBS

One of the best decisions that I ever made in connection with amateur astronomy was to join an astronomy club. In my case, it was the Toronto Centre of the Royal Astronomical Society of Canada (RASC). The society has centres in all major Canadian cities. Its 3,000 members receive several useful publications, including the indispensable *Observer's Handbook*.

Almost every city in North America with a population of more than 50,000 has an astronomy club. Members can join eclipse expeditions or participate in group observing sessions, which are excellent opportunities to seek advice and to examine a variety of telescopes. Some astronomy groups operate a club observatory built from membership funds and donations. Membership may entitle you to a discount on magazine subscriptions and astronomy books.

There are several methods of locating the nearest astronomical society. If there is a specialty store (not a department store) listed in the Yellow Pages under Telescopes, this is probably the best place to start, since such shops are in constant touch with amateur astronomers. Another approach is to call the nearest planetarium. A staff member should know whether there is an active astronomy club and where it meets. Failing that, try contacting the astronomy instructor at the local college. If these avenues prove fruitless, there may not be an astronomy club in your area, but as a last resort, call the library or regional newspaper, both of which are constantly involved with clubs and organizations of all types.

In the United States, there is no national astronomical society comparable to the RASC, although the Astronomical Society of the Pacific, an 8,000-strong group of amateur and professional astronomers in the western states, is a regional organization similar to the RASC. The Astronomical League, a quasi-national organization, is mainly an alliance of clubs, rather than an organization of individual amateur astronomers. Organizations such as the American Association of Variable Star Observers (AAVSO), 25 Birch Street, Cambridge, MA 02138, are global in membership and scope but devoted to specialized segments of amateur astronomy.

ASTRONOMY CONVENTIONS

Each year, in about a dozen locations across North America, hundreds of amateur astronomers gather for annual conventions to share their hobby, exchange ideas and look through each other's telescopes. Experienced practitioners give talks and workshops, a famous guest speaker entertains, telescope dealers and manufacturers display their product lines, and everyone has fun. Conventions have blossomed into a major element of the hobby. Some of them are extremely well organized, offering a unique opportunity to meet a wide range of aficionados.

Most astronomy conventions are held during the summer months; all are listed several months in advance in *Astronomy* and *Sky & Telescope* magazines along with addresses for more information. The biggest and/or best ones are: The Texas Star Party, held at a ranch in southwest Texas; Stellafane, near Springfield, Vermont; Riverside Telescope Makers Conference, Big Bear, California; Astrofest, central Illinois; Winter Star Party, Florida Keys (February); Syracuse Summer Seminar, Vesper New York; Starfest, Mount Forest, Ontario; Mount Kobau Star Party, British Columbia. Most astronomy conventions are held over two or three days in a location where the skies are good for celestial viewing. If one is within driving distance, don't miss it.

PLANETARIUMS

Almost every major city in North America has a public planetarium that stages "sky shows" under a huge projection dome by creating incredibly realistic images of the night sky, as seen at any time on any day from any place on Earth. In the

1930s, when the first big Zeiss planetariums were installed in New York, Chicago and Philadelphia, the shows caused a sensation. Every large city had to have one.

Today, the novelty is gone. Few new planetariums have been built since the 1970s. Some have been severely squeezed by budget cuts, and unfortunately, their shows reflect it. A positive planetarium experience, especially for a youngster, can last a lifetime. Because most planetariums produce their own shows, the quality varies from genuinely thrilling to appalling. The problem is, once you have seen a boring show, you may not have the desire to go back to *any* planetarium. If this happens, try another planetarium. Sooner or later, you will see how it should be done. Or, if your local planetarium offers quality productions, support it by seeing every new show (normally three to six a year).

Some planetariums offer introductory astronomy courses that are well worth taking. Planetariums usually have excellent telescopes available for use during such courses. A planetarium bookstore is often the best source in town for books, charts and reference material. In many cases, the local astronomy club holds its meetings at the planetarium. If the planetarium staff is doing its job, the building should be the centre of astronomical activity for the city and surrounding area.

OBSERVATORIES

Since a visit to a major astronomical observatory was a pivotal event in my infatuation with things cosmic, I strongly recommend such an excursion to any budding astronomy enthusiast.

Not all observatories have the time or the staff to offer visitors a peek through their telescopes. The main ones that do are Lick Observatory, Mount Hamilton, California, 19 miles east of San Jose (408-274-5062); Leander McCormick Observatory, Charlottesville, Virginia (804-924-7494); Allegheny Observatory, Pittsburgh, Pennsylvania (412-321-2400); David Dunlap Observatory, Richmond Hill, Ontario, 15 miles north of Toronto (416-884-2112); Dominion Astrophysical Observatory, Victoria, British Columbia (604-388-3157); Dearborn Observatory, Northwestern University, Evanston, Illinois (312-492-7650); McDonald Observatory, Fort Davis, Texas (915-426-3263); Griffith Observatory, Los Angeles, California (213-664-1181); Foothill College Observatory, Los Altos Hills, California (415-948-8590); and Chabot Observatory, Oakland, California (415-531-4560). All of these observatories have specific hours set aside for visitors. Be sure to call for details well in advance. On cloudy nights, a slide or movie programme is often available in addition to tours of the facility.

The 200-inch Hale telescope on Mount Palomar, east of San Diego, California, is the largest one in the world open to tourists. Thirteen storeys high and enclosed by a massive rotating dome, the instrument is awesome. The astronomer on duty rides in a small cage inside the telescope tube, like the commander of an interstellar spaceship. For research reasons, views of the 200-inch telescope are restricted to a glass-enclosed visitors' gallery during specified daytime hours (call 619-742-3476 for visitors' information).

Although it has no schedule of night viewing for visitors, Kitt Peak National Observatory (602-620-5350), 40 miles southwest of Tucson, Arizona, is unforgettable. During the day (10 a.m. to 4 p.m.), you can take a paved road right up the mountain. This is the largest complex of optical research telescopes in the world — 17 of them on one mountaintop. The weekend daytime tours are the best of any observatory, with movies, lectures, exhibits and impressive views of massive telescopes at an exquisite site.

Index

Anthea Weese

The Author

Terence Dickinson is Canada's leading astronomy writer. He is the author of 10 books and has written hundreds of magazine articles for publications ranging from *Reader's Digest* to *Sky & Telescope*. He also writes a weekly astronomy column for *The Toronto Star*, teaches astronomy part-time at St. Lawrence College, Kingston, Ontario, and is an astronomy commentator for "Quirks and Quarks," CBC-Radio's weekly science programme. Before turning to science writing full-time in 1976, he was editor of *Astronomy* magazine and was an instructor at several science museums and planetariums in Canada and the United States. In 1992, he received the Royal Canadian Institute's Sandford Fleming Medal for achievements in advancing public understanding of science. He and his wife Susan live near the village of Yarker in rural eastern Ontario.

Photography and Illustration Credits

Chapter 1: p.7 Gary E. Mechler; p.8 Richard Sanderson; p.9 (clockwise from top) Roy E. Gephart, NASA, Alan Dyer.

Chapter 2: p.11 through 16 Adolf Schaller; p.18 Leo Henzl, 8-inch Newtonian; p.19 Terence Dickinson, 50mm lens; p.20-21 Adolf Schaller.

Chapter 3: p.24 (top) Alan Dyer, (bottom) Terence Dickinson; p.25 Victor Costanzo; p.27 Terence Dickinson, 24mm lens; p.28 Terence Dickinson, 24mm lens; p.31 Terence Dickinson, 180mm lens.

Chapter 4: p.35 Astrophoto Laboratory; p.37, 38-39 & 40-41 Victor Costanzo; p.42 (both) Terence Dickinson; p.44-45 Victor Costanzo; p.46 (both) Tony Hallas, 5-inch refractor; p.47 Tony Hallas; p.48-49 Victor Costanzo; p.50 Terence Dickinson, 90mm lens; p.51 (both) Terence Dickinson, 28mm lens; p.52 Terence Dickinson, 28mm lens.

Chapter 5: p.55 Alan Dyer; p.56, 57 & 58 (all) Terence Dickinson; p.59 Victor Costanzo; p.60 (left) Todd Storms, (right) James V. Scotti; p.61 & 62 (all) Terence Dickinson; p.63 Randy Attwood; p.64 & 65 (all) Terence Dickinson; p.67 Jerry Lodriguss; p.69 Leo Henzl.

Chapter 6: p.71 Jim Riffle, 12-inch Astromak; p.72 Jim Riffle, 12-inch Astromak; p.73 Jack Newton, 20-inch Newtonian reflector; p.75 Michael Watson, 8-inch Schmidt camera; p.77 Jim Riffle, 12-inch Astromak; p.78 Klaus Brasch, 10-inch Schmidt-Cassegrain; p.79 Tony Hallas, 5-inch refractor; p.80 (top) John Sanford, (bottom) Alan Dyer, 180mm lens; p.82 Tony Hallas, 5-inch refractor; p.83 Bill Iburg, 14-inch Schmidt-Cassegrain; p.85 Evered Kreimer, 12-inch Newtonian; p.86 Victor Costanzo; p.87 (M101) Tony Hallas, (M81/82) Astrophoto Laboratory, (M53) Jack Newton, (4565) Jack Newton, (M6/7) Alan Dyer, (M27) Jack Newton, (Albireo) R.C. Dickinson, (M1) Jack Newton, (M35) Terence Dickinson; p.88-107 Star Charts, base photography by Ray Villard.

Chapter 7: p.109 NASA; p.110 (left) Victor Costanzo, (right) Terence Dickinson, 200mm lens; p.111 & 112 Terence Dickinson, 50mm lens; p.113 (top) NASA Jet Propulsion Laboratory, (bottom) Victor Costanzo from author's pencil sketch; p.114 NASA Jet Propulsion Laboratory; p.115 Doug Gegen, Roper Mtn. Observatory; p.116 John Bianchi from author's pencil sketch; p.117 notebook by Matthew Sinacola; p.119 NASA Jet Propulsion Laboratory; p.120 Paul Doherty; p.121 Terence Dickinson, 85mm lens.

Chapter 8: p.123 Terence Dickinson, 7-inch refractor; p.124 Terence Dickinson, 4-inch refractor; p.125 Terence Dickinson, 5½-inch refractor; p.126 Roy Bishop/the Royal Astronomical Society of Canada; p.127 Terence Dickinson; p.128 (top) Wolfgang Lille, 7-inch refractor, (centre and bottom) Terence Dickinson; p.129 (top) John Hicks, (bottom) Terence Dickinson; p.131 ©John Shaw.

Chapter 9: p.133 Alan Dyer, 3½-inch Maksutov-Cassegrain; p.134 Alfred Lilge, 12-inch reflector; p.135 Terence Dickinson, 300mm lens; p.136 John Bianchi; p.138 Richard Keen.

Chapter 10: p.141 John Sanford; p.142 (left) Terence Dickinson, 50mm lens, (right) Michael Watson, 8-inch Schmidt camera; p.144 Victor Costanzo; p.146 & 147 (top) Alan Dyer, 28mm lens; p.147 (bottom) Terence Dickinson, 24mm lens.

Chapter 11: p.149, 150 & 151 Terence Dickinson; p.152 Gary Woodcock; p.153 Terence Dickinson.

Chapter 12: p.157 Pleiades and moon with Earthshine, Terence Dickinson, 4-inch refractor.